Small Towns Resilience and Heritage Commodification

Peter Lang

Bruxelles · Bern · Berlin · New York · Oxford · Wien

Luďa Klusáková and
Blanca Del Espino Hidalgo (eds.)

Small Town Resilience and Heritage Commodification

Europe of Cultures / Dialogues
Vol. 23

This book has been published with the financial support of project KREAS: Creativity and Adaptability as Conditions for the Success of Europe in an Interrelated World (European Regional Development Fund-Project, No. CZ.02.1.01/0.0/0.0/16_019/0000734). https://kreas.ff.cuni.cz/en/

Cover picture: A view of the small town of Telč, Czech Republic, 2016. Richard Mortel via Flickr under CC Attribution 2.0 Generic (CC BY 2.0) https://creativecommons.org/licenses/by/2.0/

EUROPEAN UNION
European Structural and Investment Funds
Operational Programme Research,
Development and Education

MINISTRY OF EDUCATION,
YOUTH AND SPORTS

This publication has been peer reviewed.

Éditions scientifiques internationales
Brussels, 2021
1 avenue Maurice, B-1050 Bruxelles, Belgium
brussels@peterlang.com ; www.peterlang.com

ISBN 978-2-8076-1743-8
ePDF 978-2-8076-1744-5
ePub 978-2-8076-1745-2

ISSN 2031-3519
DOI 10.3726/b18527
D/2021/5678/47

Bibliographic information published by "Die Deutsche Nationalbibliothek".
"Die Deutsche National Bibliothek" lists this publication in the "Deutsche Nationalbibliografie"; detailed bibliographic data is available on the Internet at <http://dnb.de>.

Table of contents

Section II: Initiatives

Section III: Community

Section IV: Cooperation

Section V: Promotion

Research survey

Conclusion

Luďa Klusáková (*1950–†2020)

Jaroslav Ira

I feel honored to write this short portrait of our dear colleague and teacher, Professor Luďa Klusáková. At the same time, it is hard to accept that she will no longer be able to see this result of her work and to make so many of the plans she had in mind really happen.

Luďa Klusáková studied history and French philology at Charles University, where she later pursued her academic career as a historian of modern European history, with the main focus on urban and cultural history. For more than forty years, Luďa Klusáková was a respected scholar and gained a high reputation in international academic circles. She became professor of History and led a Seminar on General and Comparative History for two decades. Enormously active in international cooperation, Luďa played an important role in several international projects, such as the CLIOHRES.net research network (leader of thematic workgroup), TEMA Erasmus Mundus Joint Master Degree (academic coordinator for Charles University), REACH Horizon2020 project (leader of Charles University team), and KREAS research project (work-package leader). Last but not least, Luďa has long been an active member of the international committee of the European Association for Urban History (EAUH) and served as the president of EAUH from 2010 to 2012. Among her awards, one could mention the title of Chevalier de l'ordre des palmes académiques, awarded by the French Government for building relations between France and the Czech Republic in the academic sphere, and recently, honorary membership in the Finnish Academy of Sciences.

Luďa's preoccupation with cities and towns dates back to her studies, during which she focused on railways and urban spatial development. In the 1980s, she conducted a comparative analysis of the transformation of urban systems in Western and East-Central Europe on the threshold of capitalism[1]. In subsequent decades, her focus shifted to cultural themes such as identity, image of the city, and cultural encounters, and resulted in a study on the imagery and perception of the Ottoman cities in early-modern travel literature[2]. In the final phase of her career, Luďa turned her major research interest to small towns in peripheral regions of south-western France, the Czech Republic, eastern Slovakia and Poland, and analyzed them through the prism of the debates on the creative and innovative cities, revitalization strategies, uses of heritage, and urban resilience[3].

The intellectual trajectory of Luďa Klusáková represents some of the major turns in urban history: from large-scale study of urbanization and urban systems to the study of urban cultures, and more recently to domains such as heritage studies and research that borders with urban economy or tourism. Forever young in spirit, Luďa always delved into the new vocabulary with curiosity and rigor of a fresh-minded student, only to come up with an original appropriation of fashionable concepts on her long-researched themes, or, to put it another way, with fruitful rethinking of the buzzwords through her urban perspective. She was a pedant in the best sense of a historian who insisted on familiarity with wider historical contexts and intimate knowledge of the places researched, and whose awareness of the history of words used was accompanied by her openness, and indeed eagerness, to implement concepts and insights from other disciplines such as ethnology, geography, or architecture.

I still keep in my memory a course of hers that I frequented in the early 2000s. Luďa took us on a virtual tour back in time, digging

---

1 Luďa Klusáková, *Evropská města na prahu kapitalismu*, Praha 1986.

2 Luďa Klusáková, *The Road to Constantinople: sixteenth-century Ottoman town through Christian eyes*, Prague 2002; Luďa Klusáková, "Between reality and stereotype: town views of the Balkans", *Urban History*, 28, 3 (2001), pp. 358–377.

3 Luďa Klusáková, "Cultural institutions as urban innovations: the Czech lands, Poland and the eastern Baltic, 1750–1900", in: Malcolm Gee, Tim Kirk and Jill Steward (eds.), *The City in Central Europe: culture and society from 1800 to the present*, Aldershot 1999, pp. 85–100; Luďa Klusáková, "Between Urban and Rural Culture: Public Use of History and Cultural Heritage in Building Collective Identities", in: Luďa Klusáková and Laure Teulières (eds.), *Frontiers and Identities: Cities in regions and nations*, Pisa 2008, pp. 299–322; Luďa Klusáková, "Identifications of small peripheral towns at the turn of the 21st century: cultural heritage, use of historical examples, and revitalization strategies in comparative perspective", *Новое прошлое/The New Past*, «История одного города», 3 (2017), pp. 65–80.

down into the historical layers of a region in south-western France. It was an intricate exploration of many themes that Luďa so interestingly illuminated through the perspective of small towns in Gers, such as a blurred boundary between the urban and rural sphere, the uses of the past and heritage in self-promotion and rural renewal, the interplays of local, regional and national identities, and the relations between landscape, technologies and everyday culture. It was a fascinating intellectual journey that opened hearts of many of us to the themes researched in this volume. May this volume be a discreet memorial to her inspiring personality.

Foreword

Jaroslav Ira

Literature on urban resilience has recently been accumulating at an amazing pace. Yet small towns have remained at the margins of this debate, although they provide a unique perspective on resilience and may help us broaden its understanding. Small towns embody a peculiar socio-spatial category. While closely connected to the rural sphere, they represent outposts of the urban world and often proudly display their urbanity. They thus have to cope with a specific mix of challenges, different from those of the big cities and the countryside, while sharing many problems of both. We may likewise presume that the small-town communities perceive their problems and vulnerability in particular ways and often come up with original solutions. Furthermore, the histories of small towns are often fascinating testimony to the social and cultural energy that has been invested in their endurance and development. Seemingly predestined to become obsolete in modernity, small towns have largely survived as a locus of human attachment, even if their fates have varied from thriving to serious decline. While bemoaned as being in constant danger of disappearance, local identity has often served as an important basis of community resilience.

The authors of this volume offer novel perspectives on these issues. In particular, they concentrate on the role of cultural heritage in small-town resilience. It has become a truism that cultural heritage is an important source of community resilience; or that it should be, especially in smaller and peripheral places. But this idea is far from self-evident. As readers will find out, the volume brings together a range of constellations and scenarios, drawn from many different contexts. There are various reasons behind this: Many actors may have many visions. For instance, the perspective of local citizens who dramatize their fate in terms of "survival" and a "stay or go" dilemma may often differ from the perspective of planners and policy-makers who articulate the small-town question as an "economic problem", "development issue", "social question" and the like. What is considered heritage has always been a contentious issue, and so are the objectives and impacts of its use. Many projects and strategies may succeed and set the town on a trajectory to a bright future, but others may turn out to be failures, immediately or in the long run. As is often the case with social and historical enquiry, it is complexity which is the major result and message of this book.

This volume originated as a result of the stimulating exchange of papers in the session "When Local Wishes to Become Global: Heritagization Strategies of Small Towns and Little Places in Remote Regions (20th and 21st centuries)" chaired by Luďa Klusáková and Blanca del Espino Hidalgo during the conference *Urban Renewal and Resilience: Cities in Comparative Perspective*, organized in Rome by the European Association of Urban History (EAUH) in late August 2018. But its origins go back to the EAUH conference in Lisbon (2014) and the session on small towns organized by Luďa Klusáková, Marie-Vic Ozouf-Marignier and Jaroslav Ira that raised some of the issues discussed here, and that resulted in the book *Small Towns in Europe in the 20th and 21st Centuries: heritage and development strategies* (Prague 2017) edited by Luďa Klusáková. It was in Portugal that the fruitful collaboration between Luďa Klusáková and Blanca del Espino Hidalgo began. And it could be sustained thanks to the projects that supported the subsequent meetings and the publication of the present volume.

The editors are particularly grateful to project KREAS: *Creativity and Adaptability as Conditions for the Success of Europe in an Interrelated World*, a European Regional Development Fund-Project, held by the Faculty of Arts of Charles University, that co-financed the production of the volume[1]. KREAS has explored social adaptability and creativity in various dimensions. The research group around the late Professor Klusáková has focused on cultural heritage in historical perspective as a form of adaptation. How cultural heritage has been used as a means for making small towns and urban communities resilient and thus capable of steering their own futures has been one of the major questions.

1 Creativity and Adaptability as Conditions for the Success of Europe in an Interrelated World (European Regional Development Fund-Project, No. CZ.02.1.01/0.0/0.0/16_019/0000734). https://kreas.ff.cuni.cz/en/.

Equally important for the emergence of this volume was the project REACH: *Redesigning access to cultural heritage for a wider participation in preservation, (re)use and management of European culture*, supported by the European Commission within the funding program Horizon2020.[2] The project supported language revision and enabled the editors and some of the authors in Budapest to meet during the REACH conference *Resilient Cultural Heritage and Communities in Europe* in May 2018. Luďa Klusáková with her team at Charles University conducted the pilot on small-town cultural heritage, which had as its objective the exploration of the uses of the small-town heritage, with a specific focus on participatory approaches in heritage practice and research. Furthermore, REACH aimed at creating a large social platform of heritage experts and stakeholders and Blanca del Espino Hidalgo became one of the project's associated partners.

---

2 Re-designing access to cultural heritage for a wider participation in preservation, (re)use and management of European culture (H2020-SC6-CULT-COOP-2016-2017, ID 769827), coordinated by University of Coventry. www.reach-culture.eu.

Principles for heritage-based resilience in small towns facing the global era

Luďa Klusáková and Blanca Del Espino Hidalgo

This volume devoted to small towns in remote regions and the strategies they use to attract the attention of a wider, in the best case, global, public, stems from a session held at the International Conference of the European Association for Urban History in Rome, 2018. The authors of this book are convinced that, regardless of their quantity, small places remain in the shadow of big cities and metropolises, although they represent a distinct feature in the image of European culture. As such they deserve extensive coverage in research. Whereas large urban sprawls, with their concomitant suburban malls, highways, skyscrapers, and the like, may seem to lack individuality when compared globally, small towns, on the contrary, retain specific spatial organization and forms of social life, including face-to-face contact, and engender a variety of regional types[1]. If we disregard demographic and economic criteria, small towns are classified as those that maintain their traditions and cultural identity, as opposed to large towns where the focus is on the future with, in consequence, a progressive loss of the direct bond to the past. The first are categorized as "societies of memory", the second as "societies of change"[2]. Such an interpretation, however, suggests a rather gloomy prognosis for the future of small towns; on the other hand, we find that there is rooted in small communities the mobilization capacity and social and psychological importance of their tangible and intangible heritage, as well as the ability to create territorial networks on a creative and innovative basis[3].

---

1 Clark 2009.

2 Hervieu-Léger 2006.

3 Knox and Mayer 2009.

Furthermore, local cultural heritage has been pointed out as a basis from which to draw not only resources but also inspiration from how past civilizations have adapted to changing conditions; that is, have been resilient[4]. Moreover, the heritage assets – both material and immaterial – we have inherited from the past to the present day have been maintained by societies over time: therefore, we can consider them as fully sustainable.

A conceptual basis for a study about resilience in small towns and cultural heritage commodification

In recent years there has nevertheless been increasing talk of the concept of resilience in the field of urban and territorial policies as a way of implementing measures to adapt to climate change[5]. It was initially applied to urban studies concerning the adaptation of territories following the occurrence of natural disasters and the assessment of risks associated with them[6]. Nevertheless, in recent times it has been transferred to the very different academic and political spheres. Consequently, when trying to define resilience, we should not search for a single meaning but, on the contrary, for different ways of its manifestation and implementation depending on the context and the subject matter[7].

Therefore, for the subject matter of this paper, we will try to focus on the meanings and applications of the concept of resilience that have a consolidated academic corpus and that, in turn, have a relationship with the cultural heritage of small towns. In this sense, studies related to the physical or ecological aspects of small-town heritage are related, not exclusively, but as a priority, to the resilience of rural areas. Furthermore, we should not forget the incorporation of community value into the strategies and outcomes of heritage-based resilience itself, so the definition of resilient communities, particularly for small settlements, should also be addressed. That is the approach of the so-called theory of resilience thinking[8], that "addresses the dynamics and development of complex

4 Hawkes 2001.

5 Lang 2010:16.

6 Bosher and Coaffee 2008.

7 Anderson 2015.

8 Folke et al. 2010.

social-ecological systems" (SES) based on three main pillars: resilience as persistence, adaptability as the capacity to adjust the responses to external drivers and internal processes, and transformability to create new stability registers for development, all of them considered in a multiscale way. In this sense, resilience would embody the ability of an SES "to continually change and adapt yet to remain with critical thresholds".

Deepening the abovementioned implications on the resilience of adaptability and transformation abilities, cultural heritage has been identified as a vector for transformation and, in the end, for embracing change towards cultural resilience[9]. This theory is based on the fact that both cultural and natural heritage embodies what has remained after changes and crisis periods – in other words, it could adapt to change and develop through creativity and transformations – what can be reinforced by the idea that cultural heritage is always a growing and not shrinking resource.

Beyond this, a trend of particular academic relevance in recent years has been the study of community resilience, in which two branches of the traditional studies regarding resilience merge[10]: on the one hand, the ecological, based on the biophysical sciences, and on the other hand, the psychological about personal development and mental health, but from an individual approach. The community resilience incorporates their common focus on the adaptability of a social system but also incorporates matters related to the social sciences such as organization, governance, value construction or sense of place.

Cultural heritage is in fact showing a great potential when dealing with community resilience building and maintenance, particularly in small towns and rural areas, as collective activities related to memories and local symbols have contributed, throughout bottom-up volunteering processes to reinforce a collective identity[11]. This has contributed to secondary advantages such as the anchoring of the local population or, more directly, the conservation of cultural heritage assets and collections.

In this regard, we must not forget the growing role that digital networks are playing in creating and maintaining rural community ties

---

9 Holtorf 2018.

10 Berkes and Ross 2013.

11 Beel et al. 2017.

in diverse ways[12]: enabling the creation of digital tools and promoting community activities related to heritage, or reinforcing the feelings of belonging of both the present inhabitants and those who have migrated to other places through social networks. Digital tools have also enabled the linking of local rural initiatives for networking, which has offered advantages but still needs to reduce the differences in network access between urban and rural environments[13].

Finally, there has been no lack of critical voices about the generalized use of the concept of resilience from the political sphere and, especially, the excessive bias to an approach based on economic development. According to Mackinnon and Driscoll[14], in these cases, resilience would be generally defined by official guidelines, not by local communities.

However, cultural heritage-based local development has not been traditionally considered as a priority by decision-makers[15]. It has been generally seen as a resource for economic development by means of cultural tourism, which usually left small towns off the trails despite their potential[16].

On the contrary, community-based everyday appropriation of heritage has been revealed as a key for resilience[17] and, without avoiding the possibilities of encouraging sustainable tourism as a way of revitalization for small towns[18], grass roots initiatives for management of heritage resources have generally had a more balanced impact[19]. In all cases, loss of authenticity linked to heritage commodification has been pointed out as a threat[20,21].

As a result, small places today are often aware of the unique value of their tangible heritage, and attempt to play with it, with more or less success, in their development strategies. The same role may attain the

12 Roberts et al. 2017.

13 Philip et al. 2017.

14 Mackinnon and Driscoll 2012.

15 Ashworth 2013.

16 Grimwade and Carter 2000.

17 Jones and Mean 2010.

18 Prideaux and Kininmont 1999.

19 Keitumetse 2014.

20 Cohen 1988.

21 Prideaux 2003.

intangible heritage linked to the territory or the inherited goods that have not been traditionally assessed but embody their ties to the land, like minor or vernacular architecture, agricultural heritage or transition landscapes. What are the reasons? Cultural policies biased to big cities, fragmented heritage practices or absence of research on small towns? Shortage of resources, limited capacities and knowledge in small towns and peripheral regions?

A diversity of reflections, strategies and case studies for a complex and transversal issue

The objective of the session on whose base this collective book has been structured was to categorize the variety of representational strategies of the multi-level value of the heritage of small towns, including the intangible heritage and traditional economic practices, to explore links among heritage experts and other relevant stakeholders on the national and transnational level with actors in local communities, regional bodies and cross border projects, to highlight the overlooked and to identify good and bad practices, and bottlenecks.

The themes and questions we initially formulated for the session became also the core of the chapters in the book. We suggested to our co-authors that they focus on small places that are undervalued, not fully exploited, or in danger because of lack of appreciation from the core region (suffering from lack of investment in preservation, natural disasters in less populated areas, the impact of large infrastructural works – dams, tunnels, bridges, etc.). We were particularly interested in how these small places chose strategies or entering the path of revitalization, how they reach a stabilized situation, or if they fall into stagnation, decline, desertification – and what is the possible role of heritage in such various situations. Particularly interesting are the capacities of small towns to collaborate through twinning and creating projects with towns in a similar situation. Therefore, we have proposed to explore the networking initiatives of small towns that share their heritage characterization, or have a common cultural identity (cultural tourism in small destinations, cross-border initiatives, creative small cities networks, strategic planning based on heritage assets, etc.).

In the following fourteen chapters we can explore with their authors the case studies they have chosen for this book. They confirm our hypothesis that the large towns and metropolises show more elements, making their typology easier, while the small towns perform such a rich variety of types, that the typology becomes rather difficult. Hélène Mainet[22] based her simple dual typology of small towns on French experience: the first type are small towns on the edges of metropolitan zones offering their inhabitants, on the one hand, proximity to sophisticated urban life and, on the other, access to the calm pleasures of nature around them. The second type consists of small towns far from metropolises, deep in the countryside[23], which offer their inhabitants at least basic urban comfort, culture, and administrative services, even creating centres of regional importance, although small[24]. The Polish scholar Andrzej Sztando[25] concluded that very important for the success of small town strategies are the supralocal contacts quite necessary for the development of networking. Both approaches are meaningful; however, we have chosen as the leading principle in our book the ways to enhance commodification of the heritage to strengthen the resilience of the community. Among other concepts devoted to express the ability of small towns to face the challenges of a globalized and polarized current urban system[26], we decided to highlight the paradigm of resilience, understood as the ability of an element or organism (originally nature and the human being or, more recently, local communities) to adapt itself to changes and return to a point of equilibrium after the usual environment and resources have been transformed or eliminated. This has been more recently linked to the kind of processes that have affected inhabited places and, more precisely, to the urban heritage fabric[27].

Going deeper into the contributions, the views of the different authors on the processes are diverse, as are the strategies followed by the small towns studied for the commodification of their heritage resources. Thus, one of the major conclusions derived from a comprehensive reading of the proposals is the revelation of social capital as a key factor, that is, the capacity of the inhabitants of these small towns

---

22 Mainet 2011.

23 Storey 2006.

24 Mainet op. cit.

25 Sztando 2017.

26 Harvey 1989.

27 D'Amico y Currà 2014.

to generate heritage commodification processes, to recover old sustainable traditional practices and, even more, to maintain them over time as an effective resource that supports resilience. This point can be particularly noted in chapter 2 (Does the developmental history have an influence on small town's thinking? by Réka Horeczki), chapter 7 (A New Phoenix from the Ashes? by Marta Marçal Gonçalves et al.,) and chapter 8 (Building community resilience through heritage in small towns: a case study of Vysoké nad Jizerou in the 1990s, by Jiří Janáč). A common feature of the heritage characterization of some of the case studies has also been the existence of a significant manufacturing industry which became obsolete, then was transformed into a cultural resource for social development, heritage commodification and resilience. That is the case with the agricultural landscape and architecture in chapter 3 (Tourism and territorial development: dynamization strategies for an agrarian cultural landscape. Guadalhorce valley in Málaga, by Lourdes Royo Naranjo) and the industrial heritage in chapter 5 (The white coal town. The emergence and decline of the textile industry in Naoussa, by Nikolaos Leonidakis). Large scale joint strategies among groups of small towns have also been considered in several contributions, particularly about landscape units holding natural heritage in chapter 9 (Corredor Bio-Comechingones, an innovative regional scale agreement among small towns, by Monica Jimena Rame), and railway infrastructures in chapter 11 (Railway: from global infrastructure to local heritage. The case of the historic railroad Avellino-Rocchetta in Southern Italy offered by Consuelo Isabel Astrella). In other cases, as in chapter 2 from Réka Horeczki (Does the development history influence how we think about small towns?) and chapter 10 (Heritagization and networking in some extra-small north Italian places), by Paola Pellegrini, networking initiatives of small settlements with a territorial or cultural common history have been incorporated, both including experiences in cross-border regions. In other examples, the interest of the case studies has been found in the resilience of small towns after dramatic events such as a huge forest fire in chapter 7 or the abandonment of a village due to the construction of a reservoir in chapter 6 (Ruin and time, abandoned settlements. The small town of Granadilla contributed by Francisco Javier Ostos Prieto). Actual projects for heritage enhancement from a planning perspective as in chapter 4 (Viterbo_heritage and future, by Manuela Raitano) have also been incorporated. Readers will be able as well to find several mentions of the consequences of cultural tourism policies on the urban and social management of small towns, with a critical point of view on certain occasions as in chapter 14 (Touristification of historic urban centres in the Southern Italian Salento region, by Katja Maaria Huovinen). Finally, external factors have been revealed as crucial in some contributions as in the case of chapter 1 (Can small towns survive in a global world?, by Lars Nilsson), with an emphasis on UNESCO World Heritage declarations and how they have put remote small towns into a global perspective, as can be seen in chapter 12 (Balancing Between Local and Global: Heritage Presentations of Central European Small Towns. The Case of Telč and Bardejov, by Jan Krajíček) and chapter 13 (Entry into the UNESCO club: the experience of the Ulyanovsk region, offered by Elena Elts).

Regarding the spatial distribution of the case studies, we are presenting a collective book with a territorial balance, mainly centred on the European context (as the proceedings of a conference of the European Association for Urban History), including Spain (chapters 3 and 6), Italy (chapters 4, 10, 11 and 14), Portugal (chapter 7), Greece (chapter 5), Sweden (chapter 1), Hungary and Romania (chapter 2), touching cross border collaboration with Slovenia (chapter 10), and the Czech Republic (chapters 8 and 12) compared with a case in Slovakia. Nevertheless, we wished to open the field of study to other locations outside the European Union. Thus, our contributors offered two interesting cases showing how globally important is this issue in articles devoted to a case in Argentina (chapter 9) and another in Russia (chapter 13). We found this view beyond the borders quite refreshing and inspirational and hope that the readers of our book will appreciate this as well.

Thus, the authors provide a multi-faceted, comprehensive vision of the potentiality but also of the complexity of making small towns resilient through heritage commodification processes. Diversity is guaranteed through a wide range of geographical locations, scales of approach, multidisciplinary authorship, critical or enthusiastic points of view, priority granted to local/external actors or citizens, cultural identities, and consideration of heritage resources with a variety of nature. Nevertheless, two fundamental issues are raised as threads throughout this collective work. On the one hand, the fact that small towns continue to acquire increasing relevance in the study of urban history and social, demographic, cultural, political and heritage processes. On the other hand, the revelation that heritage is becoming more and more a concept not limited to the study of the past but, fundamentally, a valuable resource when the communities of small towns wish to solve current problems, and hopefully, a guide towards a better world.

Bibliography

Anderson, B. (2015). What kind of thing is resilience? *Politics*, 35(1), 60–66.

Ashworth, G. J. (2013). Heritage and local development: A reluctant relationship. In *Handbook on the Economics of Cultural Heritage* (pp. 367–385). Edward Elgar.

Beel, D. E., Wallace, C. D., Webster, G., Nguyen, H., Tait, E., Macleod, M., & Mellish, C. (2017). Cultural resilience: The production of rural community heritage, digital archives and the role of volunteers. *Journal of Rural Studies*, 54, 459–468.

Berkes, F., & Ross, H. (2013). Community resilience: Toward an integrated approach. *Society & Natural Resources*, 26(1), 5–10.

Bosher, L., & Coaffee, J. (2008). Editorial: International perspective on urban resilience. *Urban Design and Planning*, 161, 145–146.

Clark, P. (2009). *European Cities and Towns (400 – 2000)*. Oxford.

Cohen, E. (1988). Authenticity and commoditization in tourism. *Annals of Tourism Research*, 15(3), 371–386.

D'Amico, A., & Currà, E. (2014). The role of urban built heritage in qualify and quantify resilience. Specific issues in the Mediterranean city. *Procedia Economics and Finance*, 18, 181–189.

Folke, C., Carpenter, S. R., Walker, B., Scheffer, M., Chapin, T., & Rockström, J. (2010). Resilience thinking: Integrating resilience, adaptability and transformability. *Ecology and Society*, 15(4), 20.

Grimwade, G., & Carter, B. (2000). Managing small heritage sites with interpretation and community involvement. *International Journal of Heritage Studies*, 6(1), 33–48.

Harvey, D. (1989). From managerialism to entrepreneurialism: The transformation in urban governance in late capitalism. *Geografiska Annaler: Series B, Human Geography*, 71(1), 3–17.

Hawkes, J. (2001). *The Fourth Pillar of Sustainability: Culture's Essential Role in Public Planning*. Common Ground.

Hervieu-Léger, D. (2006). In search of certainties: The paradoxes of religiosity in societies of high modernity. *The Hedgehog Review*, 8(1–2), 59–69.

Holtorf, C. (2018). Embracing change: How cultural resilience is increased through cultural heritage. *World Archaeology*, 50(4), 639–650.

Jones, S., & Mean, M. (2010). *Resilient Places. Character and Community in Everyday Heritage*. Demos.

Keitumetse, S. (2014). Cultural resources as sustainability enablers: Towards a community-based cultural heritage resource management. *Sustainability*, 6, 70–85.

Knox, P. L., & Mayer, H. (2009). *Small Towns Sustainability: Economic, Social, and Environmental Innovation*. Birkhäuser.

Lang, T. (2010). Urban resilience and new institutional theory: A happy couple for urban and regional studies. In Müller, B. (ed.) *German Annual of Spatial Research and Policy 2010* (pp. 15–24). Springer.

Mackinnon, D., & Driscoll Derickson, K. (2012). From resilience to resourcefulness: A critique of resilience policy and activism. *Progress in Human Geography*, 37(2), 253–270.

Mainet, H. (2011). Les petites villes françaises en quête d'identité. Ambiguïté du positionnement ou image tactiquement combinée? *Mots. Les langages du politique*, 97, 75–89.

Philip, L., Cottrill, C., Farrington, J., Williams, F., & Ashmore, F. (2017). The digital divide: Patterns, policy and scenarios for connecting the 'final few' in rural communities across Great Britain. *Journal of Rural Studies*, 54, 286–398.

Prideaux, B. (2003). Commodifying heritage: Loss of authenticity and meaning or an appropriate response to difficult circumstances? *International Journal of Tourism Sciences*, 3(1), 1–15.

Prideaux, B., & Kininmont, J. (1999). Tourism and heritage are not strangers: A study of opportunities for rural heritage. *Journal of Travel Research*, 7(3), 299–303.

Roberts, E., Beel, D., Philip, L., & Townsend, L. (2017). Rural resilience in a digital society: Editorial. *Journal of Rural Studies*, 54, 355–359.

Storey, D. (2006). Images of rurality: Commodification and place promotion. In *The Rural Citizen: Governance, Culture and Wellbeing in the 21st Century*. University of Plymouth.

Sztando A. (2017). *Ponadlokalna perspektywa zarzadzania strategicznego rozwojem lokalnym na przykladzie malych miast* [Supralocal Perspective in Local Strategic Development Management on the Example of Small Towns] (pp. 110–120). Wydawnictwo Uniwersytetu Ekonomicznego we Wroclawiu – Wroclaw Economic University Publishers.

Section I

Reflections

Can small towns survive in a global world? The cases of Avesta and Söderhamn

Lars Nilsson

Introduction

Globalisation and the transition to a post-industrial economy at the end of the twentieth century were in many European nations followed by an increasing number of shrinking cities. In fact, declining cities were sometimes much more frequent than growing cities. Many of those cities and towns that lost inhabitants were rather small in population size, but even medium-sized and major cities could have serious growth problems. Besides, shrinking cities had often an industrial background, and their economy had mostly been dominated by one or a few branches and companies. The urban demographic and economic expansion had generally started during the nineteenth century industrialisation and with some interruptions continued up to de-industrialisation in the 1960s and 1970s. From a geographical point of view, shrinking cities were spread all over the nation but with concentration in certain regions, which have been called rustbelts[1].

Another new trend by the late twentieth century was metropolisation. At the same time as a majority of all cities and towns shrank, metropolitan areas and major cities grew quickly. This growth process started often after a period of metropolitan decline caused among other factors by the so-called green wave. Many people preferred in the 1970s to flee the major cities and settle down in small towns or on the rural countryside. The urban-rural population balance stabilised at a high level, and the space for further increase of the urban degree was only marginal.

---

1 For an overview see Nilsson (ed.) 2011.

Post-industrial metropolitan growth has demographically been caused by urban to urban migration. Net migration streams have gone from small towns to major cities and not, as previously, from rural to urban areas. Therefore, the urbanisation rate has remained rather constant. Other important sources for post-industrial metropolitan growth have been immigration, not least of refugees, and a high net-birth rate.

These new tendencies mean that we may talk about a post-urban world. The urban degree is so high that it cannot increase much more, and most population changes take place inside the urban system. Urban norms and values are dominating even on the rural countryside.

Fig. 1. Map of Sweden including case studies' location. Source: Own elaboration.

In this article I will discuss the possibilities for small towns to survive in a global and post-industrial world with examples from two Swedish shrinking towns, Avesta and Söderhamn (Fig. 1). They are located not so very far from each other, 200–300 kilometres north of Stockholm, in what we can call the Swedish rustbelt (see map). Their decline started with the de-industrialisation in the 1970s. The local authorities have since then used a wide variety of measures to regain growth. Today, both municipalities have around 20,000–25,000 inhabitants, including both urban and rural areas (Fig. 2).

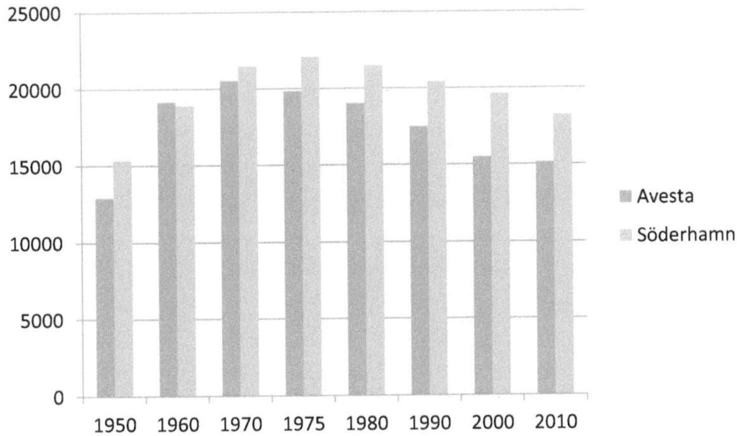

Fig. 2. Number of inhabitants for the urban regions of Avesta and Söderhamn, 1950–2010. Note: An urban region includes the city and its suburbs, and follows here the geographical limits from 2010.
Source: Nilsson 2011a, www.scb.se/Statistikdatabasen.

Two shrinking towns

Avesta

Avesta is located along the river Dalälven in the interior of the region Bergslagen, well-known for its mines and big manufacturing companies. It was the combination of waterfalls, and the rich supply of copper ore, iron ore and woods, that made this site suitable for industrial production. Avesta became in the seventeenth century the main centre for Swedish copper production. Copper ore was delivered by horse transports from the copper mine in Falun, 68 kilometres north of Avesta, and charcoal from farmers in the surroundings. Smiths were recruited both from Sweden and other nations. The copper production became so important that Avesta was granted privileges as a town in 1641. Avesta also acquired a monopoly for the production of coins. The factory may during its heyday have been the largest in the world of its kind.[2]

The town privileges were revoked in 1686, copper production failed and Avesta stagnated and became a shrinking town. A new era began in the nineteenth century based on iron and steel production. The international demand for Swedish iron and steel increased substantially during industrialisation, and the iron factory in Avesta, established in 1821, developed quickly after the Johnson family became sole owners in 1906. The number of inhabitants rose in a few years from 2,000 to 3,600.

The expansion was also stimulated by the construction of the railway system from Stockholm to Northern Sweden. Avesta was originally intended to be a central node in the Swedish Northern railway system, but due to geo-technical problems the railway station had to be located in a neighbouring place. In any case, the advent of the railway also had positive effects on Avesta. Population increased up to 5,000 in 1919, when Avesta re-acquired its status as a town.

During the economic crisis of the 1920s and the 1930s Avesta became once again a shrinking town. This lasted until the late 1930s before the population figure was back at the same level as in 1919. The rebuilding of Europe after World War II meant a very high demand for Swedish iron and steel. Avesta Iron Works increased production and employment and kept its position as one of the leading Swedish steel producers. There were also other big factories in the neighbourhood of Avesta on the other shore of the river Dalälven, for example an aluminium company, a chlorate factory and a phosphate industry.

The population of Avesta increased significantly during these fruitful years. One obstacle to further growth was the scarcity of land. Huge tracts of land were owned by the Iron Works, and not for sale. In any case, the number of inhabitants amounted to 20,000 in the late 1960s, suburbs included.

---

2 This section of Avesta is based on Nilsson and Båve 2016: 26 ff. See also: Avesta 1958, Avesta 2002, Norberg 1956. Population figures from Nilsson 1992, 2011a. Web pages: https://sv.wikipedia.org/wiki/Avestakoncernen; https://sv.wikipedia.org/wiki/Outokumpu; https://sv.wikipedia.org/wiki/Outokumpu_Abp; https://sv.wikipedia.org/wiki/Krylbo.

Increasing international competition and rising oil prices caused serious growth problems for the entire Swedish steel industry in the 1970s. The branch was totally reorganised, including among other things rationalisation of production and fusion into fewer and larger units. Avesta Iron Works survived this process, but acquired new owners in several stages ending up with the Finnish State company Outokumpu.

Avesta Iron Works is still today an important enterprise in Avesta, but with fewer employees than before. The aluminium factory, on the other side of the river Dalälven, closed down in 1975, the chlorate and phosphate companies even earlier. This de-industrialisation process resulted immediately in population decline. The number of inhabitants in Avesta urban region fell from 20,000 in 1970 to ca 15,000 in 2010. This gives in relative terms a reduction of 25 per cent in forty years.

Söderhamn

Town privileges for Söderhamn were issued in 1620 due to the establishment of an armaments factory. Armaments forging was an old specialism in this region, and through the privileges the production was concentrated in the town. Only six other Swedish towns had such factories during the seventeenth century, when Sweden was a great power in European politics. Production fell in the eighteenth century, and in 1813 the factory was definitively closed down. The burghers had to rely on the traditional urban trades, such as handicraft, commerce and fishing[3].

A new epoch started in the mid-nineteenth century with the sawmill industry, whose initial expansion was financed by capital from Stockholm and Göteborg. This was on the eve of Sweden's industrialisation, and Söderhamn had a favourable geographical location as a port city on the southern coast line of the Gulf of Bothnia. The rich stock of forests could easily be transported on the river Ljusnan from the interior of the region to Söderhamn at the coast. Several steam-driven sawmills were established around Söderhamn, together with for example timber

---

3 This section of Söderhamn is based on Nilsson and Båve 2016: 45 ff. See also Grandien 1926, Jörnmark 2007. Population figures from Nilsson 1992, 2011a. Web pages: https://sv.wikipedia.org/wiki/Arizona_Chemical; https://sv.wikipedia.org/wiki/Bergvik_och_Ala; https://sv.wikipedia.org/wiki/Ljusne-Woxna AB; https://sv.wikipedia.org/wiki/Söderhamn.

yards. The final products were to a great extent exported, and shipped to national customers as well. The economic base was, as for Avesta, much wider than the local and regional market in itself. Population size was determined by the city's function as an export node.

The sawmill expansion had a lot of positive external effects on other trades. An engineering industry started for example, for the production of machines and equipment for the sawmill industry and other companies. Infrastructural investments in roads, railways, waterways, harbours, residential houses etcetera became necessary as well. The economy flourished, and the number of inhabitants in Söderhamn rose from less than 2,000 persons in the mid-nineteenth century to more than 10,000 in 1890. Besides, many new urban localities developed around the factories outside the city's administrative area.

The sawmill production reached its peak at the end of the century. Some factories closed down because of increasing international competition from Finland and Russia. Besides, the exploitation of the Swedish forests had been so intense that the raw materials for the sawmill industry became a scarce resource. Production was concentrated on fewer and bigger units. The population and the economy stagnated. The sawmill decline was however counterbalanced by an upswing for the paper pulp industry which stabilised the situation. In any case, Söderhamn was still strongly dependent on forest industries.

The great depression in the 1930s had serious consequences for Söderhamn. Only 8 out of 23 sawmills survived. The paper pulp industry also had severe problems and the effects on the other trades were wide-ranging. The population declined to below 10,000 inhabitants. The situation did not improve until after the Second World War. The recovery became possible because of two big investments, one by the Swedish Government, and the other by the phone company Ericsson. The Swedish State decided in 1945 to localise an armaments wing to Söderhamn, and Ericsson invested shortly afterwards in a new production unit. This resulted in a more differentiated economy but Söderhamn was still strongly dependent on manufacturing, and not least forest industries. The economy flourished again, and the number of inhabitants increased to 20,000 in 1970, suburbs included.

The strong dependency on manufacturing meant that Söderhamn once more came into a precarious situation, when deindustrialisation began in Sweden. The first signs of an up-coming crisis became visible in the mid-1970s. Several pulp factories reduced production and fired workers. In the next ten years more than one thousand industrial jobs were lost, due to a mix of reduced production and rationalisations. The number of lost jobs was the same for the following ten years. Many factories could not survive during the new post-industrial conditions. Initially, the forest industries had the greatest economic problems, but even the phone fabric and other engineering industries reduced production and fired people.

In the 1990s the Swedish Parliament voted for a new defence policy. The number of military units should generally be reduced, and the military division in Söderhamn was closed down in 1998. A few years later the phone company Ericsson sold its works in Söderhamn to the American company Emerson. They stooped production in 2004 and liquidated the factory. As we remember, the armaments wing and the phone factory saved Söderhamn from the former economic crises; now they were both gone.

In spite of the deindustrialisation, manufacturing is still an important sector in Söderhamn, and comparatively more strongly represented than in most other Swedish cities. One of the largest companies today is Arizona Chemical with its head office in Florida. About 100 workers are engaged in the distillation of raw pine oil in a suburb of Söderhamn, Sandarne. This is the largest unit in the world for pine oil production. Arizona Chemical is also an example of how global capital and multinational companies have gradually taken over the industries, and the owners are not present in the city as they were when industrialisation began.

Local measures against population decline

External help

During industrial times, Swedish cities facing economic crises usually relied on help from the Government. A combination of State measures and private investments in new factories by leading manufacturing companies had been a successful strategy in the 1950s and the 1960s. It was seen as a national interest that the State and the private capital should work together to solve local crises on the labour market. Only the

State had the necessary overview to decide how to use scarce resources in the best way. Municipalities were therefore by law forbidden to sponsor private companies, and compete with each other[4].

Initially, the local authorities of both Avesta and Söderhamn hoped for external help to handle the upcoming industrial crisis in the 1970s. The local authorities of Söderhamn, for example, expected that private entrepreneurs should now realise their previous plans of building a new large-scale paper mill close to the city. Thousands of people demonstrated and claimed support from the State. Local delegations were sent to Stockholm to argue for the need of external money to save jobs by setting up new production units to replace the losses[5].

Investments in a paper mill were of course not at all realistic in a situation with extensive de-industrialisation, and the State could no longer assist shrinking cities in the same manner as before. Instead, local authorities were urged to solve their own problems without external subsidises. All previous restrictions regarding municipal sponsoring of private companies, place marketing, and local economic policies were abolished.

Local economic policies

During these new circumstances many Swedish municipalities, including Avesta and Söderhamn, began almost immediately with active place marketing and sponsoring of private businesses. Local economic policies were formulated and decided by the city councils. The local administration was strengthened by special offices for economy and commerce. By such measures, the lost industrial jobs were supposed to be replaced by new industrial jobs. It was a local industrial replacement strategy.

The local authorities of Söderhamn pursued a very active local economic policy with a focus on small- and medium-sized firms. A number of companies owned by the city itself were set up to promote private investments and attract new ones. The municipality even took over a chain factory that was threatened by bankruptcy. Shortly afterwards this

---

4 Nilsson and Forsell 2013: 245 ff. See also Henning 1996, Olsson1995.

5 Lindberg 2002: 104–128.

factory had in any case to close down and be liquidated with substantial economic losses for the city[6].

The local authorities of Avesta used similar strategies to their colleagues in Söderhamn. A municipal foundation, later reorganised as a municipal company, was founded to attract and support new investments. Potential investors could lease the necessary premises, and get loans on favourable conditions. These measures resulted in the establishment of a number of new small and medium-sized firms. But the newcomers were only short-lived. Within a few years they closed down or left Avesta, and were sometimes replaced by new short-lived companies[7].

This industrial replacement policy gave very meagre results in both cities. The labour market did not improve. Population decline continued. It turned out to be more or less impossible to replace lost industrial jobs with new jobs of the same kind. The economic crisis deepened and was followed by deficits in the municipal budget. This financial crisis was caused by diminishing tax revenues and increasing municipal costs. Avesta tried to improve the public finances by selling off municipal property, for example the energy company and housing foundations[8].

The local authorities of Söderhamn handled the excess of housing in another way. Instead of selling off municipal property they decided to pull down the surplus of tenement houses. Such actions could be subsidised by the State[9]. The State also assisted Söderhamn with a special "package", with focus on IT and education, to compensate for the loss of the military wing[10].

Culture and tourism

When the industrial replacement strategy failed, the local authorities choose a new course directed towards culture and tourism. Both culture and tourism had been expansive branches in the post-industrial Swedish economy. The new strategy seemed therefore to have good potential for regaining population growth. The loss of manufacturing occupations

---

6 Ibid. pp. 152–161. http://soderhamnsandan.se.

7 Nilsson and Båve, op. cit.: 58 ff.

8 Nilsson and Båve, op. cit.: 66 ff.

9 Nilsson and Båve op. cit.: 79 ff.

10 Lindberg op. cit.: 62.

should be replaced by investments in culture, cultural heritage, tourism, and services.

A former industrial area in Avesta, named the Northern Works, was in the 1990s rebuilt as a cultural centre. One of the old buildings became site for an international exhibition – Avesta Art – of contemporary art. Another building was renovated as an industrial museum recalling Avesta's long industrial history. Later on, the exhibition centre and the museum were supplemented by an art academy. The Northern Works also includes a sports arena and small industrial firms. The area has been marketed as The Copper Valley with reference to the previous copper production from the seventeenth century[11].

The art exhibition, Avesta Art, became a great success and since the start in 1995 has attracted thousands of visitors every year. In 2010 the number of visitors exceeded 20,000 which was almost as many as the total population of the municipality.

Avesta's long tradition of mining and manufacturing has not, however, been totally forgotten. A rising demand for minerals has given great expectations for a flourishing future based on the region's wide-ranging knowledge of mining. Several industrial projects have been initiated in cooperation with other municipalities, regional authorities and university colleges[12].

The local authorities of Söderhamn gradually changed strategy in the same way as their colleagues in Avesta when all efforts to re-industrialise the city failed, as well as the state package. Nevertheless, in the 1980s still, urban planning gave priority to restoring the old industrial structure. However, the ambition to return to the same number of inhabitants as before the industrial crisis gradually weakened and finally seen as unrealistic. Shrinking was more or less accepted and the goal became to stabilise the economy and the population on a lower level than before. Urban planning from the mid-1990s emphasises therefore tourism, cultural heritage, museums and services, instead of large-scale industrial projects[13].

---

11 Nilsson and Båve op. cit.: 65 ff. See also Geijerstam af (red.) 2007: 128, Storm 2005. Web pages: https:://sv.wikipedia.org/wiki/Koppardalen; http://www.verket.se.

12 http://www.jernkontoret.se; http://www.triplesteelix.se; http://www.avesta.se/2020

13 Söderhamn, Plan 1995.

City renewal

The local authorities of both cities have, as well as all their other actions, used urban renewal as a means to make the city more attractive for potential visitors, investors, and inhabitants. Several proposals for restructuring the inner city were presented to the city council of Avesta in the 1960s. The planning was accomplished in close collaboration with local merchants. Because of uncertain industrial forecasts the local authorities wanted to rebuild and market Avesta as a commercial centre for the entire region. A property company partly owned by the city was established in 1969 to facilitate the restructuring process. The new city centre was opened a few years later (1973) with a pathway through the city centre and parking on the outskirts[14] .

A new main road for through traffic was built outside the city. Previously all cars, buses and lorries had to pass through central Avesta irrespective of their final destination. An external shopping centre was established in the 1980s at a traffic hub along the new main road.

City planning was updated several times later on, for example, in 2010. A new travel centre was then proposed, combining buses and trains with a pathway to the Copper Valley. It was presumed that the Copper Valley's accessibility as well as its attractiveness should increase in that way.

The local authorities of Söderhamn tried in the same manner to strengthen the city's role as a regional trade centre. In the early 1990s an external shopping centre was opened along the motorway, E 4, outside Söderhamn with all the facilities such a centre was supposed to offer[15]. The need for a modernisation of the inner city was emphasised in a city plan from 1998. A careful restructuring and densification of the central quarters was proposed and decided. The same goals were underlined in the 2015 city plan, and formulated as "a living city centre with a feeling of archipelago"[16].

The efforts to encourage entrepreneurship and innovations survived in projects such as "Drivkraft", "Högtrycket" and "Ungdomslyftet". They emanated from the program "Söderhamn Vision 2012", which had

---

14 Nilsson and Båve op. cit.: 52 ff.

15 http://www.e-center.se

16 Söderhamn, Plan 1998, 2015.

education and diversity as a fundamental pillar. Besides, the projects had the ambition to create a positive image of Söderhamn, and in that way make the city attractive[17].

A new railway station in the outskirts of central Söderhamn was opened in 1997 and closely linked to local buses. The city's accessibility and attractiveness was supposed to increase by linking long-distance and short-distance travelling. The old centrally placed railway station became a cultural heritage and tourist goal. The new station resulted in an increase of commuting, especially out-commuting, from Söderhamn[18].

Summing up

The urban local authorities of Söderhamn and Avesta have thus by a number of different measures tried to regain population growth after decades of decline. In the first stage they hoped for external help to re-industrialise the cities. Those efforts were however not successful. In the next step the local authorities tried to re-industrialise their towns by own means such as local economic policies, sponsoring of private businesses and place marketing. Later on, they turned to culture and tourism in combination with programs for urban renewal. Furthermore, they re-structured the towns towards regional commercial centres. But all measures were in vain. The number of inhabitants continued to decline.

A new population trend after 30–40 years of decline?

Population statistics for the period 2010–2015 reveals that the number of inhabitants has increased in Avesta as well as in Söderhamn. But not only in these two previously shrinking cities. In fact, most Swedish cities registered population increases and the number of shrinking cities returned to the same low level as in the 1950s and the 1960s. A "normal" level of shrinking cities was about one out of three in late industrialism[19].

Avesta's population increased by 1.58 per cent per year in the period 2010–2015, which was far above the national average of 1.05 per cent.

---

17 Leffler and Falk-Lundqvist 2014; Vallström 2011. Web pages: http://www.söderhamn.se/vision2012; http://soderhamnsandan.se.

18 Nilsson and Båve op. cit.: 72–74.

19 Nilsson 2018.

Suddenly, Avesta became one of the more expansive Swedish cities. The annual population growth of Söderhamn was more modest, or 0.49 per cent (Fig. 3).

The demographic explanation for these new traits is immigration, and foremost immigration of refugees. Both cities had since long ago registered net-immigration, but this surplus of foreign migrants was not enough to cover for losses by net out-migration to other Swedish municipalities and net deficits of births. This situation changed dramatically in the years 2010–2015. The number of net-immigrants increased so much that it covered not only population losses by net domestic out-migration but also population decline because of more deaths than births (Tab. 1). Avesta even acquired a small surplus of domestic in-migration.

The labour market did however not expand very much. Söderhamn had an increase of employment of only two per cent in the period 2010–2015, and Avesta had lower employment in 2015 than in 2010. The influx of refugees made it possible for Avesta and many other cities as well to grow demographically even if the labour market did not improve at all. This cannot be a sustainable growth in the long run.

Even if the labour market did not expand there were important structural changes. Avesta saw a re-structuring from manufacturing mainly towards education, health care, medical care, and social care. These branches were also generally very expansive in Sweden during this period, especially among cites that like Avesta turned from stagnation to population growth. Söderhamn, on the other hand, totally missed these expansive branches. Instead, there was a re-structuring from manufacturing towards the less expansive branches of public administration, commerce and cultural services. This may explain the differences in population growth between Avesta and Söderhamn in 2010–2015[20].

It is of course impossible to know today if these new tendencies will sustain in the long run. We must wait until we have data sets for longer time spans than five years.

---

20 Ibid.

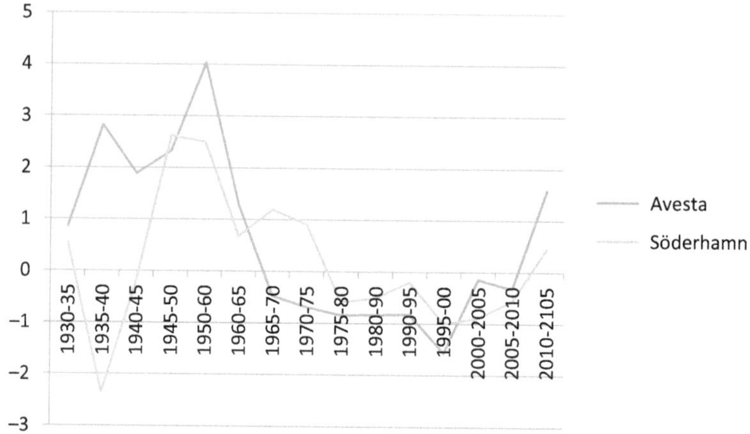

Fig. 3. Annual population growth rates for the urban regions af Avesta and Söderhamn respectively, 1930–2015. Source: Nilsson 2011a. www.scb.se/Statistikdatabasen.

Tab. 1. Population changes in the municipalities of Avesta and Söderhamn, 2001–2015. Source: Own elaboration based in data from www.scb.se/Statistikdatabasen.

	Net births	Net immigration	Net domestic migration	Population change
Avesta				
2001–2005	-666	262	-6	-410
2006–2010	-436	628	-554	-554
2011–2015	-367	1493	53	1179
Söderhamn				
2001–2005	-708	233	-691	-1166
2006–2010	-695	1241	-1384	-838
2011–2015	-492	1802	-1174	136

Can small cities survive? Conclusion

Cities normally grew in waves. Periods of expansion are followed by contraction and then a phase of growth may start again. Economic crises have for centuries been a normal part of urban life in Avesta as well as in

Söderhamn. Both cities have recovered mainly thanks to external forces and not by own actions.

The recent growth problems have in the same manner as previous crises been brought to a halt at least temporarily by external forces far outside the control of the local authorities. Political, military, and economic conflicts in northern Africa and the Middle East have sent hundreds of refugees to such remote places as Avesta and Söderhamn. The arrival of refugees has broken the falling population trend. This new upswing is however not the consequence of a flourishing labour market as previous population growth periods were. Therefore, it is uncertain if the growth can sustain in the long run. Hopefully, the newcomers will contribute with impetus to the local economy and a sustainable expansion.

In this article I have tried to demonstrate that measures from local authorities have not been enough for a shrinking city to regain population growth. This is not to say that local policies are of no value at all. They may very well soften the negative effects of economic stagnation even if they cannot cope with the latent structural mechanisms behind the decline. But to regain population growth, external forces seem to be necessary.

Urban decline is of course not a new phenomenon. During each decade of the industrial era about one third of all Swedish urban regions lost inhabitants. However, these losses did not normally last very long. Stagnation occurred generally without population loss.

Urban decline has been much deeper and more frequent since the start of the post-industrial and global era. Many urban regions had therefore fewer inhabitants in 2010 compared to the 1970s. This leads to the question: can urban decline continue forever, if there is no external stimulus? My answer is: probably not as long as the city upholds the role of a central place. A central place needs a minimum number of inhabitants to be able to deliver services to its own citizens and to those living in the surroundings. Sooner or later a minimum point will be reached, population decline may stop, and the number of inhabitants will thereafter be more stable.

Central place duties can of course also be reduced. Centralisation of, for example, education, health care, banking, commerce, and other services can affect many towns in a negative way. The supply of services of various kinds diminishes, and then, the minimum population level for service distribution will be lowered. The town can be downgraded to a lower central place level, and the risk for shrinking increases. The digital revolution may as well reduce central place functions and be a threat towards the survival of small towns.

Besides being central places Avesta and Söderhamn were also export nodes. Their population sizes were determined by this latter function. De-industrialisation meant that the export sector needed fewer people and resources than before which of course also resulted in lower demand for central services. The balance between central place functions and export functions changed in favour of the former. Population size had to be adjusted to these new circumstances, which meant shrinkage.

After 30–40 years of decline Avesta and Söderhamn had perhaps reached, or were close to, that minimum population level necessary for fulfilling their roles as central places. If so, decline would not have continued much longer. Besides, they still maintained export functions. In any case, immigration of refugees must have been the most important factor behind their recovery in 2010–2015. Thus, as previously, external factors were decisive for the regaining of population growth.

Swedish urban development can be described in long waves with a length of about forty years. The periodisation has been as follows: c. 1840 – 1870s/1880s – 1920/1930s – c. 1970 – c. 2010. Each period has had its own dynamic branches which have favoured some cities and disfavoured others[21]. It is of course still impossible to know whether a new epoch, with fewer shrinking cities and new prime motors, started around 2010.

Bibliography

Avesta. (1958). *Avesta jernverk. Dess historia, anläggningar och produkter.* Avesta.

Avesta. (2002). *Avesta jernverk. Historien om hur böndernas smedja blev en världskoncern.* Avesta.

Geijerstam af, J. (red). (2007). *Industriarv i förändring.* Avesta.

Grandien, A. (1926). Söderhamn. In *Sveriges städer, vol. 5.* Stockholm.

Henning, R. (1996). *Att följa trenden. Aktiva och passiva kommuner.* Stockholm.

---

[21] Nilsson 2011b: 171–251.

Jörnmark, J. (2007). *Övergivna platser.* Stockholm.

Leffler, E., & Falk-Lundqvist, Å. (2014). *Drivkraft Söderhamn*, Umeå.

Lindberg, H. (2002). *Att möta krisen: Politikbyte på lokal nivå i Söderhamn 1975–1985.* Uppsala.

Nilsson, L. (1992). *Folkmängden i administrativa tätorter 1800–1970.* Stockholm.

Nilsson, L. (2011a). *Folkmängden i tätortsregioner 1950–2005.* Stockholm.

Nilsson, L. (2011b). Urban Development in Sweden, 1950–2005. In L. Nilsson (ed.), *The coming of the post-industrial city: Challenges and responses in Western European urban development since 1950* (pp. 171–251).

Nilsson, L. (ed.) (2011c). *The coming of the post-industrial city: Challenges and responses in Western European urban development since 1950.* Stockholm.

Nilsson, L. (2018). Ett nytt urbant skede? Unpublished paper.

Nilsson, L., & Båve, E. (2016). *Krympande orter: Avesta och Söderhamn som postindustriella samhällen.* Stockholm.

Nilsson, L., & Forsell, H. (2013). *150 år av självstyrelse: Kommuner och landsting i förändring.* Stockholm.

Norberg, P. (1956). *Avesta under kopparbrukets tid, vol. 1–2.* Stockholm.

Olsson, J. (1995). *Den lokala näringspolitikens lokala ekonomi: En jämförande kommunstudie.* Örebro.

Söderhamn, Plan (1995). Fördjupad översiktsplan för Ljusnans mynning. Antagen av kommunfullmäktige 1995-08-28. Söderhamns kommun.

Söderhamn, Plan (1998). Fördjupad översiktsplan för Söderhamns stadskärna. Antagen av kommunfullmäktige 1998-12-21. Söderhamns kommun.

Söderhamn, Plan (2015). Utveckling av Söderhamns centrum. Planprogram maj 2015. Söderhamns kommun.

Storm, A. (2005). *Koppardalen: Om historiens plats i omvandlingen av ett industriområde.* Stockholm.

Vallström, M. (2011). *Att återerövra framtiden.* Lund.

Web pages

https://sv.wikipedia.org/wiki/Arizona_Chemical

https://sv.wikipedia.org/wiki/Avestakoncernen

https://sv.wikipedia.org/wiki/Bergvik_och_Ala
https://sv.wikipedia.org/wiki/Koppardalen
https://sv.wikipedia.org/wiki/Krylbo
https://sv.wikipedia.org/wiki/Ljusne-WoxnaAB
https://sv.wikipedia.org/wiki/Outokumpu
https://sv.wikipedia.org/wiki/Outokumpu_Abp
https://sv.wikipedia.org/wiki/Söderhamn
http://www.avesta.se/2020
http://www.jernkontoret.se
http://www.scb.se/Statistikdatabasen
http://www.triplesteelix.se
http://www.verket.se

Does the development history influence how we think about small towns?

Réka Horeczki

Introduction

Small towns have recently been playing a more important role in local development than ever in Hungary. Most Hungarian and East European urban researchers prefer analysing big cities and agglomerations due to the accessibility and multitude of information. Correspondingly, the study of small towns can have a supplementary function among the theories. This study is based mainly on Hungarian academic discourse and aims to present the long-term developmental roles and characteristic features of small towns, which entities strongly influence the economic, social and political life of Eastern and Central Europe. The number of small towns in this area, especially in Hungary, has permanently changed during the last centuries (with market town development, the urban laws of 1871), yet their proportion increased mainly after the regime change in the early 1990s and the Millennium (with a proliferation of large villages with town status). My paper applies a comparative method; it investigates primarily the historical development and opportunities of countries with a significant small town network disposing of similar assets and structures. For example, the South Transdanubia region in Hungary, the Centru Development Region in Romania, Moravia in the Czech Republic, and "*Felvidék*" in Slovak Republic – with many small towns – are among the twenty most disadvantaged regions of the European Union, which reflects a significant decline compared to previous years. These regions were border regions and peripheral in the Austro-Hungarian Empire, too. This, in my opinion, justifies the need for professional research directed towards changing this situation through the tangible results of the local actors' efforts. The investigation of the small towns of these areas and the drafting of their possible development path in harmony with existing development plans (economic and cultural) may provide a positive vision of the future not only for the small towns, but their wider environment as well. The factors which have relevant impacts on small towns' development potentials are their geographic position, historical development, and the emergence of civil society. Beside these aspects, local traditions, practices, special age-long agricultural activities, the region's representations, specialised knowledge or economic attitude (craft industries, special horticultural activities) are also relevant features. One hidden solution for the small towns may be the creation of social capital, which can be used to improve the local quality of life.

Settlement system in Eastern Europe

The settlement network as a whole is a stable system, where most changes occur in the long term. The changing political systems, alliances and state borders were sufficient to bring about considerable changes in the settlement network, especially in the case of capitals and large cities[1]. In the 1990s, the more or less homogenised system of settlements was affected to a significant degree by phenomena such as deindustrialisation, de-urbanisation, privatisation, foreign capital investments and the possibility of local autonomy, which led to increased competition between different hierarchical levels and elements of the settlement system[2]. New factors of success became decisive in the location choice of economic actors. The emerging process of transformation, considering old structures, produced significant disparities in the system of settlements and strengthened spatial concentration, which had especially relevant consequences for differences in competitiveness. The economic space of Eastern Europe is dominated by urban areas, leading to a top-down diffusion of modernisation and innovation from the highest levels of the settlement hierarchy[3]. Considering that due to their size and geographical position, large cities and small towns fulfil the role of economic, employment and service centres in their regions with a population of a million, in the framework of our research it is reasonable to analyse

---

1 Hajdú 2014.
2 Enyedi 1998.
3 Horváth 2000, 2015.

regional functions fulfilled by urban areas along with the conditions of competition.

One of the most important issues of the European Union's regional policy nowadays is the development of urban and metropolitan areas. It is generally accepted that urbanisation is one of the most powerful, permanent forces in the world. We live in a new urban era with most of humanity now living in towns and cities. Over the 50 years from 1951 to 2011, population growth in the EU was at its highest in the 1960s when the increase was 8 per cent over the decade. Growth slowed down gradually up to the 1990s to an increase of around 2 per cent in the decade. Growth in the 1960s was mostly concentrated in the urban regions, where there was an increase of 12 per cent over the decade, as against 9 per cent in intermediate regions and 1 per cent in rural regions[4]. After 1971, differences between population growth in the EU-15 and the EU-13 became more marked. Growth in urban and intermediate regions was slightly above average, while in rural regions it was around half the average. In the EU-13, population growth slowed down after the 1980s and became negative after 1991. In all three types of regions the population fell during the 1990s and continued to fall in rural regions. These changes in population growth were accompanied by changes in the degree of urbanisation in the EU. Compared with the rest of the world, the EU, especially the EU-15, was already highly urbanised in 1961. In the 50 years since then, the proportion of the population in the EU-15 living in cities has not changed. Between the 1960s and the 1990s, the population living in towns and suburbs increased and the proportion in rural areas fell from 30 per cent to 25 per cent. In the EU-13, the degree of urbanisation is lower.

4 Dijkstra–Poelman 2014.

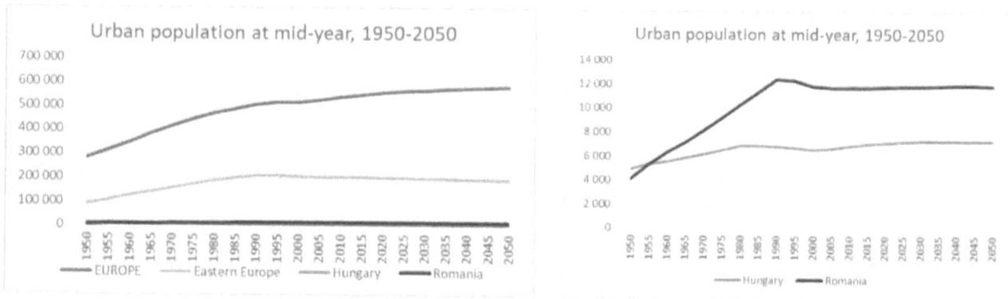

Fig. 1. Urban population at mid-year, 1950–2050. Estimation from 2020 (thousand people). Source: author's editing based on OECD urbanisation data. OECD urbanisation data available: https.//data.oecd.org/.

Hungarian and Romanian urbanisation processes (Fig. 1) have progressed on similar lines from 1990. The main differences can be observed between their population size and the dynamics before the regime change. Currently, there are 346 settlements with town status in Hungary (See Fig. 2). Seven-tenths of Hungary's population lives in the towns; two-thirds of the towns (182) have a population under 30,000. Half the towns have a population of less than 8,634.

Why is it important to analyse Hungarian small towns?

Towns have always had an important role in the settlement network of Europe. The development of towns and the issue of inequalities among towns have always been important and accurate topics. Urbanisation processes became stronger after the 1990s and have formed specific trends in the development of settlement networks. Within Europe, urban development in Central and Eastern European countries shows specific development patterns. An important feature respecting the settlement network is that big cities, mostly capitals, have become dominant, and counterpole cities do not really exist.

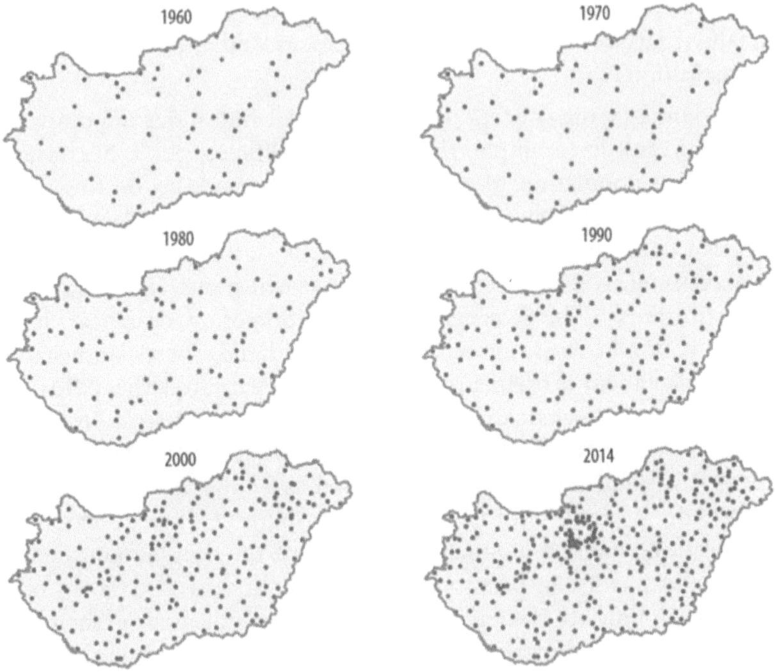

Fig. 2. The change in Hungarian towns between 1960 and 2014.
Source: author's editing based CSO 2015.

Baranya county has 10 small towns (out of 22), Somogy county has 11 small towns (out of 16)[5]. Most of these towns have important natural endowments, but not all of them are able to utilise them. The quality of settlements which have town status is quite diverse regarding their population and infrastructural and functional characteristics. Towns are the power centres of economic life due to their population density, heterogeneous employment build-up, and because their morphologic image and standards are different from those of villages.

According to the literature, the definition of the town can be derived from multiple components. Towns can be defined from the aspects of statistics, economics, geography, architecture, economic history, society, etc. According to the definition in economics, a town is a settlement

5 CSO 2019.

that serves as a centre for its surrounding villages through providing specialised services beyond everyday needs, as well as its administration and institutions[6].

The demographic concept usually sets a minimum size of population to define a city; for example, the Statistical Office of the UN considers areas with a population of more than 5,000 people to be towns (see Tab. 1). The most widely accepted definition among Central and Eastern European countries is the one that concerns status, because towns are those settlements "that have administrative town status, rights, authority and institutions"[7]. The diversity of the definition of towns appears in the context of small towns, as well. On the one hand, there is a negative, critical voice about settlements that became towns after the 1990s[8]; on the other, small towns can have a meaning that suggests a "forgotten category"[9], while others claim that they do not appear at their deserved place[10]. Partly, for a village being declared as a town is an opportunity for upgrading; along with this, it is the tribute of their development. Classification based on population seems to be the simplest and the easiest to follow, as it can be supported by exact statistical data. In other words, the definition of small towns is not only indicated by their population, though it can be categorised with the help of it, it is also based on several other criteria. According to several authors (see Tab. 1), small towns are settlements that had town status in 2011, whose population was generally under 10,000 people between 1870 and 2011, which received city status after 1989/90, and within whose population there exists a social group that can be considered as the advocate of town development, a well-established bourgeois class; and through their history the places were settlements serving central functions.

6 Enyedi 2012, Szigeti 2002.
7 Illés 2008: 140.
8 Beluszky–Győri 2006.
9 Bácskai 2002: 12.
10 T. Mérey 1989.

Tab. 1. The definition of small towns by population criteria. Source: Own editing.

Author	*Definition*	*Criterion*
Heineberg (2014)	Small town	5,000–20,000 inhabitants
Zuzanska-Zysko (2005)	Small town	Below 20,000 inhabitants
Vaishar (2004)	Small town	Under 15,000 inhabitants
Tóth (2008)	Small town	Under 5,000 inhabitants
	"Small-middletown"	5,000–10,000 inhabitants
Dövényi (2009)	Small town	5,000–15,000 inhabitants
	"Tinytown" or "Minitown"	Under 5,000 inhabitants
Hungarian National Territorial Development Plan	Small town	Under 30,000 inhabitants
Administrative Organisation of Romanian Territory (2005)	Small town	5,000–30,000 inhab. + 10–20 km commuting zone

Taking the historic antecedents into account, the settlements of Hungary reached nineteenth century conditions through several changes, but the specific features of their development started to appear in the period afterwards. At the beginning of the nineteenth century 743 settlements had a town status, of which 692 settlements were called "*oppidum*". According to the Hungarian Demographic Lexicon, an oppidum is a town with rights which were between those of a royal free city (*libera regia civitas*) and a village. This notion was widespread in Hungary in the fourteenth and fifteenth centuries, when hundreds of villages received market town status. In Hungary, the term oppidum referred to an economic centre with privileges, specific legal status and the right to host markets. The analyses of market towns between the two world wars were dichotomous. Mályusz thought of the role of oppida in a negative way; according to him, these settlements can be considered towns on the basis of their privileges, yet they are just rural settlements with undeserved rights[11]. However, Bácskai states that oppida have to be comparable not only with towns, but also with villages. This comparativeness enables us to study market towns in their own specific form. As a result of that, oppida based on their earlier privileges or privileges

[11] Bácskai op. cit.

received from the feudal lord gained a comprehensive self-government. In the settlement hierarchy oppida were on a level above villages, as they shared all the characteristics of villages, but they gained different privileges, among which the most common was the right to host markets. Towns were differentiated by their specific legal status, their higher population compared with villages, and their economic and infrastructural central and organising role. For this reason, it is important to examine the life of market towns and small towns through the study of their development. These towns did not have any developed public institutions or public utilities, but in some regions they emerged as centres for commerce, industry, public administration, public education or culture.

In the South Transdanubian region (Baranya, Somogy and Tolna county) there were 56 oppida with more than 2,000 inhabitants (see Tab. 2); excluding Toponár with 983 inhabitants and Bükkösd with 275 inhabitants. In Tolna county many of the oppida preserved their status and undergone a spectacular progress.

Tab. 2. Oppida in the South Transdanubian region (Census, 1828)[a]. Source: Author's editing after data provided by 1828 Census (Nagy Ludovico)[b].

Baranya county	*Somogy county*		*Tolna county*	
Bóly	Ádánd	**Kaposvár**	Báta	Pinczehely
Dárda	**Nagyatád**	Karád	**Bátaszék**	Regöly
Gödre	Babócsa	Kéthely	**Bonyhád**	**Simontornya**
Kozár Rátz	Berény Iharos	Törökkoppány	**Dombóvár**	Szakcs
Magócs	Berki Nagy	**Marcali**	Döbrököz	**Szekszárd**
Mohács	Berzence	Mernye	**Földvár**	**Tamási**
Pécsvárad	Böhönye	Mozsgó	Hőgyész	**Tolna**
SzabadSzent Király	Bükkösd	Nágocs	Iregh	
Siklós	Csokonya	Sárd	Kölesd	
Szekcső	**Csurgó**	Segesd	Ozora	
Vaiszló	Györök Szőlős	**Sellye**	**Paks**	
	Hidvég Város	**Szigetvár**		
	Igal	Toponár		
	Istvándi			

a Those oppida which are currently towns are marked in **bold**

b Available: https://www.europeana.eu/portal/hu/record/9200332/BibliographicResource_3000095461009.html

The April laws of 1848 had a significant role as they withdrew the privileges of the estates. With the change of property rights and the economic situation, the factors of settlement development changed to some extent; serf emancipation generated migration, the emerging capitalist economy boosted the different branches of transportation and facilitated the establishment of manufacturing industry. From the aspect of the settlement system, the Austro-Hungarian Compromise of 1867 can be considered a significant turning point, as it established the political requisites of the changes in the bourgeois class. After the Compromise the country aspired to implement an independent economic policy, within which infrastructural developments were carried out. This coincided with a period of economic boom; the industrial revolution motivated the formation of social and legal conditions and increased the competitiveness of regions with good agricultural conditions. The importance of the agricultural boom should be emphasised, as the change took place in Hungary, an agricultural country, where agricultural products and their trade remained the most important internal sources of capital accumulation. The modernisation of agricultural production and its technology, the expansion of trade, the modernisation of processing, and the establishment of credit institutions have all become sources of town development. There was a salient change in the nineteenth century concerning the settlement network: the share of village population started to decrease. Until World War I, the town population increased three times faster than the village population. Correspondingly, Tolna county experienced a dynamic growth of the population in its small towns. However, the actual demographic reason was not the high rate of natural increase in town population, since the number of births was significantly lower in towns than in villages, but the death rate was the reverse (due to the better public sanitation and housing conditions, infant mortality was lower, and children could reach adulthood with a better chance). The high rate of migration to towns compensated the low rate of natural increase, and villages provided cities with labour force and population[12].

Not only did population growth support the establishment of towns, but the development of public administration also had an important role. A county system organised on the basis of unified principles was created in Hungary; the counties were divided into districts, thus eliminating

---

[12] Beluszky-Győri 2005.

the areas and regions that previously had privileges. At the time of the Compromise more than 800 settlements had town status, 90 per cent of which had a market town status. No consistency is found in the labelling of towns throughout history, since after the Compromise the market town status was replaced by two new legal categories: the town with an integrated council and the town with municipal status. Towns with an integrated council can be considered the descendants of market towns from the aspect of legal status and judicial authority. Towns with an integrated council belong directly under county governance[13]. From the point of view of my research, this category seems to be the most interesting, as towns with municipal status formed separate public administration units; their population typically exceeded 20,000; they were regarded as middle or large towns. Currently existing small towns were either towns with an integrated council or large villages in this period. From 1870 to 1950, the status of large village was given to settlements that, although they had no integrated council, employed notaries.

The category of administrative centre, in which the development of towns was determined mainly by institutions and functions related to public administration, refers to differences beyond the scope of legal status[14]. Becoming an administrative centre meant that town functions were established in the settlement, including the setting up of courts, land offices and police stations. The officers working at these institutions provided the basis for the local bourgeois class; their consumption and needs increasingly represented urban attitudes[15]. After the setting up of public offices, several institutions with central functions emerged, such as schools, press and medical institutions. From the aspect of the development of small towns, the construction of railways became an important factor. The establishment of railways increased the importance of the settlements in many cases, helping them to become towns, as in the case of Dombóvár, Szigetvár or some settlements along Lake Balaton. Small towns on the south coast of Balaton were located on the line of the Southern Railway; thus, their location was auspicious, because spa-culture became one of the basic elements of bourgeois life from 1895 on, which reinforced the development of small towns. While the main

---

13 Máthé 1996.

14 Bibó 1975.

15 Kovács 1980.

lines were built more quickly in the southern counties, the building of branch lines was only realised after the turn of the century. The commuter rail lines, like Kaposvár-Fonyód, Barcs-Kaposvár or Kaposvár-Siófok, strengthened the central role of a given settlement and supported the emergence of other (sometimes public administrative) functions. The sharp difference between settlements with town status and those with town functions is the main characteristic of the Hungarian town network that prevailed before World War I. At the turn of the century, in addition to the already existing 131 settlements with town status, 250 settlements served town functions as lower-rank centres considering their number, population, economic structure, and the institutional capacity of public administration. The image of the settlement network at this time shows the predominance of the capital and the lack of small and medium-sized towns.

The settlement network of Hungary was also modified to a great extent by the Treaty of Trianon. The capital became absolutely dominant, and there were no regional centres that could have functioned as its counterpole. Not only the settlements, but also the settlement network changed. In 1918 there were 139 settlements with town status, by 1920 this number decreased to 47. The number of villages also decreased drastically: there were 12,500 beforehand, and now there are 3,500 within the new borders. The borders also altered the country's economy and transportation network, and suburbia. The town network declined and lacked important suburbs, and it had an uneven territorial distribution. The rate of town population barely changed; primarily the population of Budapest and that of its agglomeration changed[16]. The development of agrotowns in the Great Hungarian Plain was stagnating in this period, as opposed to the industrial towns on the northern part of the country. After the wars the agrotowns reached the limits of their economic performance and were unable to implement structural change. Their population growth did not stop, and thus they had to face several social inequality problems in addition to structural problems. Despite the territorial losses, the country's urban network still progressed as in the period before World War I. Essential structural problems and territorial unevenness caused serious problems, and there were still too many deficiencies in the town areas. The lower base of the urban network was

---

16 Beluszky 1999.

still missing, like the versatile developed small towns. Most of the small towns still undertook only one supply function (the role of being a transport junction, market place, educational centre etc.).

From the point of view of the settlement network we can think of 1950 as the end of a stage, like the Compromise of 1867; when the reform of public administration came into force. In addition to the transformation of the county system, town networks started to expand, and the number of the towns started to grow[17]. The first phase of urban development was the period from 1950 to 1965, when nine settlements were declared towns. Seven of them were so-called "socialist towns", they were either mining or industrial centres. The designated place for Sztálinváros seemed to be a fortunate one, as it became the centre of a town deficient region. Their primary functions were designed to serve the industrial site, while the population of the town was only secondary. At that time 106 settlements represented Hungary's small and medium-sized urban network. Large villages served similar functions as the currently existing small towns, but only half the settlements which served the function of the towns, actually belonged to towns legally. The National Concept for Settlement Network Development of 1971 (hereinafter referred to as NCSND) introduced a strict categorisation for the settlements, under which nine categories were distinguished. The towns I examined belonged to the following three categories: secondary centre, partial secondary centre, and lower priority level centre. The NCSND's focus on the city network enabled further development for those that had been designated as having an important central role[18]. Until 1980, in this category, 18 new cities were "born" which met the expectations laid down in the NCSND. The population of small towns grew at a moderate pace; industrial development improved the network of institutions and their function as regional centres strengthened. Significantly more towns, nearly 50, were inaugurated up to the mid-1980s. Many of these settlements were well developed and rich in central institutions.

By this time, the group of urban settlements that functioned as towns, caught up in number to the settlements which had civic authority. Two small towns were inaugurated in this period in Southern Transdanubia: Siklós received town status in 1977 and Tamási did so in

---

17 Germuska 2002.

18 Horeczki 2015.

1984. From 1989 there was a sharp increase in the number of the towns which were inaugurated in Hungary, namely, 41 (Tab. 3).

Tab. 3. Small towns in 2011 in the South Transdanubian region[a]. Source: Author's editing by CSO 2015.

Baranya county		Somogy county		Tolna county	
Town-declaration in					
Bóly	*1997*	Balatonboglár	1991	Bátaszék	1995
Harkány	*1999*	*Balatonföldvár*	*1992*	Dunaföldvár	1989
Kozármisleny	2007	*Balatonlelle*	*1991*	*Gyönk*	*2009*
Mágocs	*2009*	Csurgó	1989	*Nagymányok*	*2009*
Pécsvárad	*1993*	*Fonyód*	*1989*	*Simontornya*	*1995*
Sásd	*1995*	*Igal*	*2009*	Tamási	1984
Sellye	1997	*Kadarkút*	2005		
Szentlőrinc	1996	*Lengyeltóti*	1992		
Villány	2000	*Nagybajom*	2001		
		Tab	1989		
		Zamárdi	2008		

a Table 3 indicates the small towns in 2011 in the South Transdanubian region; those with fewer than 5,000 inhabitants are indicated in italics

Csurgó, Fonyód, Tab and Dunaföldvár belonged to them in South Transdanubia. In 1996, there were 206 towns in Hungary, while a discrepancy again existed between functional and formal towns; there were more settlements with town status than were needed. Several settlements received town status in 1995, such as Balatonboglár, Balatonföldvár, Simontornya, etc. These settlements received town rank again after losing market town rank, because the shape, the population, the activity of the settlement and their central function was not appropriate after the World Wars and Socialism.

After the change of regime and the development of the settlements, their population, employment structure and history differ in several aspects. Despite the high number of towns, the geographic distribution is still uneven; there are still many areas which lack the presence of towns nationwide and in regional breakdown, too. The settlement network's different structure is reflected by the spatial distribution of the towns. In the summer of 2005, 15 urban settlements received town status in the country, and the number of towns thereby increased to 289. In 2010, the number of towns increased to 304 (excluding the capital and the county seats). In many cases the declaration of settlements as towns is only the final result of a bureaucratic process; it does not expand the settlements' functions. Comparing the number of existing towns in 2005 and in 2010 we can conclude that nearly three-quarters of Hungarian towns belonged to the small town category. After 2010, the process of town declaration seems to stop; over the years since then the number of the towns has not increased in Hungary. After the town declaration boom in Pest county in 2013, the concept of the large village became obsolete. Over the years, the urban population as a proportion has increased steadily. Although the size of the urban population decreased in 2010 due to the adverse demographic wave and the process of suburbanisation, this did not shift its proportion (Fig. 3).

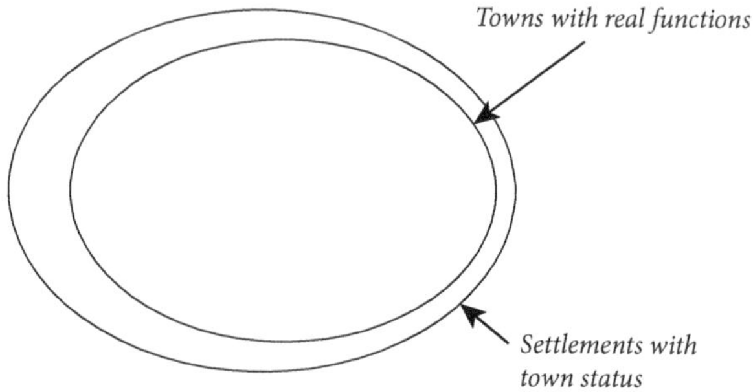

Fig. 3. The main problem with the urban system in Hungary.
Source: author's editing.

Within the urban population (Tab. 4), the proportion of the population of small towns reflects a significant concentration in some counties; on average, one-fifth of town inhabitants live in small towns. Baranya county is below the national average, which shows the superiority of the county seat, Pécs, compared to the rest of the towns in the county. Kaposvár and Szekszárd have fewer inhabitants; therefore the proportion of the population living in small towns in Somogy and Tolna counties is 5 percentage points higher than in Baranya county.

Tab. 4. Small towns in 2018 in the South Transdanubian region. Source: Author's calculation based on KSH 2018.

	Somogy county	Baranya county	Tolna county
Population (people)	303,802	363,721	219,317
Number of towns	16	14	11
Urban population (people)	158,279	238,385	123,434
Number of small towns			
Under 3,000 inhabitants	6	4	2
Under 5,000 inhabitants	9	7	3
Under 10,000 inhabitants	11	10	6
Population of small towns (people)			
Under 3,000 inhabitants	11,105	9,992	4,084
Under 5,000 inhabitants	28,295	22,254	7,949
Under 10,000 inhabitants	39,016	43,677	30,435
Proportion of small towns' population (%)			
Under 3,000 inhabitants	3.7	2.7	1.9
Under 5,000 inhabitants	9.3	6.1	3.36
Under 10,000 inhabitants	12.8	12.0	13.9

Issues about town development and urbanisation, the formal and functional concept of a town, and the conceptual and legal issues about town declarations are of great concern in the literature and are currently related mostly to small towns. Hungarian regional journals have published a number of articles related to small towns: as an incomplete functional element[19], about the types of globalisation[20], their demographic transformation[21], as well as general statistical summaries and analyses[22]. Tóth[23] initiated a debate about town declarations in 2008 and 2009 in *Területi Statisztika* relating to the inaugurated towns, which was a result of the town boom after the change of regime. Differences caused by the formal and functional duality of the concept of the town were a specific problem. Researchers approached the issue from several aspects, everyone from their own perspective. Through the example of a given settlement

19 Beluszky–Győri op. cit.

20 Molnár 2006.

21 Pirisi–Trócsányi 2006.

22 Kőszegfalvi 2008, Szebényi–Nagyváradi 2009.

23 Tóth 2008.

we can read beyond the theoretical approach to a detailed exploration of the social and legal differences and then a pragmatic approach. According to studies, small towns can be categorised most logically on the basis of two criteria: the number of inhabitants, and functionality. Classification on the basis of the population is not homogeneous, but unifying proposals were and are being made. It is important to follow these debates in the case of the small towns examined, especially in connection with the towns that were declared as towns after 2005, since they cannot fulfil the essential criterion for being a city, which is the provision of some basic functions. The majority of small towns cannot be considered as cities by their appearance, neither is their range of services appropriate, and sometimes not their infrastructure. If we examine the availability of important institutions (pharmacy, paediatrician, administrative office for citizens), two-thirds of small towns are not urbanised, or less so. Nearly 100 small towns do not have a nursery, one-fifth of small towns lack secondary educational institutions, and only one-tenth of the 185 small towns have land offices, museums and local bus services[24]. Beluszky–Győri[25] calls these small towns that became cities after the change of regime, "urban villages" and "almost towns". The authors examine the occurrences of urban settlements without the role of a city. From the aspect of economic history, functions and services that ensured the achievement of a central role (market centres and market towns, a district seat) these were what allowed these settlements to become towns after the regime change. The Parliamentary Resolution in 2005 about the NCSND already took the European Union's specifications on the terms of the settlement network development into account. The NCSND defined a territorial vision of the country, in which the relationship between the town and its region projected a new, intensive and integrated role. It was also designed to ensure the development of smaller towns to build a qualitatively renewed rural-urban relationship, where the division of functions can deepen, and the availability of high-quality centre settlements would be ensured. Creating a polycentric, cooperative urban network is one of the key objectives that would effectively link together the different regions of the country. One possible way of polycentrism is the establishment of functional urban areas. The big cities with international importance would provide the basis of the structures of these urban areas; additionally, the

24 Kovacsicsné 2010.

25 Beluszky-Győri op. cit.

basis would be the small towns with regional roles, and they could be integrated into the larger units. The goals pursued by the NCSNDgoals up until 2013 included the development of connectivity between city networks. In addition, the NCSND mostly focuses on regional development poles. It emphasised sustainability, so the plan was to design cities with developed cultural and traffic junctions. In the case of the relations between the centres and sub-centres, the emphasis was placed on ensuring accessibility.

According to VÁTI's[26] research in 2008, the main problems of the urban system are: suburbanisation, the weakness of the nodes, and the deficiencies in network cohesion and management. The scope of formal cities and small towns is one of the main sources of settlement dilemmas. Hungary's deficient urban areas "were filled", a process that is spectacular in South Transdanubia, since each South Transdanubian micro-region has at least one small town. The deficiency of urban areas ceased to exist, although towns and town-regions appeared to function with deficit. An apparent result of the emerging process of urbanisation is that the range of functional cities is narrowing, but the number of settlements with urban status has been growing since the regime change. The weak, rural small towns are not able, or only partly, to fulfil their function, i.e. to fulfil the central role in the given region. In most small towns, access to public institutions and public services is not guaranteed either, which may affect the people living in the locality negatively and damage the town image, in other respects already not positive.

Heritage in Somogy county's small towns

Generally speaking, when looking at the cultural heritage in Hungary, it can be said that there has been significant progress. The territory of the county of Somogy has been colonised for thousands of years. Reminiscences of different periods (prehistoric and "*Árpád*-era") does exist now for example, the ruin of stone castle on the periphery of Kadarkút. The architectural heritage is modest in *Somogy* county because buildings, mansions, written records suffered total destruction or irretrievable loss in Turkish period and during the World Wars.

---

[26] VÁTI mean: Hungarian Regional Development and Urbanism Public Benefit Society.

Buildings and national monuments were constructed in the eighteenth and nineteenth centuries. The historical inner town can hardly be found in Hungarian small towns, due to the different historical background/ development path[27].

The heritage of the county is part of a general statutory scheme; the "Lake Balaton" Law, the Spatial Development Perspective, the Spatial Plans and the Local Spatial Plans, the Towns Plans do include a cultural heritage chapter. These questions are primarily an economic issue and secondarily, a consideration of adequate protection and maintenance to keep them in good condition. Every small town in the county has a church that is classified as a national or local monument (Fig. 4).

Fig. 4. Churches in Nagybajom in the middle of the twentieth century. Source: Somogy County Archives, V. 276.

[27] SMÖH 2014.

There were a great many large estates in current small towns (Márffy, Somssich, Tóthi-Lengyel, Vassdényei, Festetics, Széchenyi, etc.), and there were many castles, halls and mansions in the manors of these families Several buildings are in good condition, hosting schools, social services or nursing homes. Many local architectural monuments function as "*tájház*" (cultural and social community homes), or have kept the original function (pharmacy, mill, dwelling-house – see Fig. 5.)

Fig. 5. Then and now (1850 and 2019) – Kadarkút (Somogy county). Source: Somogy County Archives, V. 261. and author's photographs.

Conclusion

Throughout the history, the low level of cooperation between the cities was a constant difficulty, which has not been solved even in the twenty-first century. Physical links between settlements (infrastructure, entrepreneurial ecosystems) are the ones that have been built in most cases and operate sustainably. Economic, social and other intermunicipal interactions do not function so well. The fragmented local government system generates a competitive situation, in which, from the point of view of the town network, small towns lose. With the centralisation of public services and public institutions, several functions, which had otherwise an important role in population retention and attraction, were removed from the scope of small towns. Among the possible development models, the expansion of the small towns' network constantly appears, with the aim of building a stable town hierarchy. The regional development model realised under the decentralisation and the renewed nationwide network model based on de-concentration are both counting on the lower elements of the urban network and urban-rural relations (Rechnitzer, 2008). Small towns represent an important part of the city network, proportionally; the goal is to express this excess in their function and appearance, and in the ability to enforce their interests. Current development policy in Hungary lacks consideration of city networks (Faragó, 2008); since the NCSND of 1971, it is not typical of development policy to create a uniformly operating concept about towns. Centralisation can be observed in development policy and in public administration, which is also reflected in the reconstitution of district centres in 2013. The NCSND accepted in 2012 did not break this tradition; at the same time, it determines the territorial approach to regional development and industrial development as well.

Acknowledgements

Project no. 132294 has been implemented with the support provided from the National Research, Development and Innovation Fund of Hungary, financed under the K-19 funding scheme. The research was partly financed by the Higher Education Institutional Excellence Programme of the Ministry of Human Capacities in Hungary, within the framework of the 4th thematic programme "Enhancing the Role of Domestic Companies in the Reindustrialisation of Hungary" of the

University of Pécs (reference number of the contract: 20765-3/2018/ FEKUTSTRAT).

Bibliography

Bácskai, V. (2002). *Városok Magyarországon az iparosodás előtt.* Osiris kiadó, Budapest.

Beluszky, P. (1999). *Magyarország településföldrajza. Általános rész.* Dialóg Campus kiadó, Budapest-Pécs.

Beluszky, P., & Győri, R. (2005). *Magyar városhálózat a 20. század elején.* Dialóg Campus, Budapest–Pécs.

Beluszky, P., & Győri, R. (2006). Ez a falu város! (Avagy a városi rang adományozásának gyakorlata s következményei 1990 után). *Tér és Társadalom*, 2, 65–81.

Bibó, I. (1975). *Közigazgatási területrendezés és az 1971. évi településhálózat fejlesztési koncepció.* Magyar Tudományos Akadémia, Budapest.

Central Statistical Office. (2015). *Hungarian CSO Yearbook, 2014.*

Central Statistical Office. (2019). *Gazetteer of Hungary 2018.*

Dijkstra, L., & Poelman, H. (2014). A harmonised definition of cities and rural areas. The new degree of urbanisation. Regional Working Paper. European Commission.

Dövényi, Z. (2009). „Város az, ami magát annak nevezi". Tűnődések Tóth József tanulmánya kapcsán. *Területi Statisztika*, 1, 3–7.

Enyedi, Gy. (1998). *Transformation in Central European Postsocialist Cities.* Pécs, Centre for Regional Studies, Discussion Papers, 21.

Enyedi, Gy. (2012). *Városi világ.* Akadémiai kiadó, Budapest Falu Város Régió. 3.

Faragó, L. (2008). A funkcionális városi térségekre alapozott településhálózat-fejlesztés normatív koncepciója. *Falu Város Régió*, 3, 27–31.

Germuska, P. (2002). A szocialista városok létrehozása. *Századvég*, 2, 49–73.

Hajdú, Z. (2014). A városba tódulás és a városfejlődés a dualizmus időszakában Magyarországon. In Z. Takács, A. Ricz (Eds.), *Regionális kaleidoszkóp.* Szabadka. Regionális Tudományi Társaság, 81–90.

Heineberg, H. (2014). *Stadtgeographie.* UTB.

Horeczki, R. (2015). Small town-question in the scientific literature. In P. Nijkamp, M. Buček (Eds.), *5th Central European Conference in Regional Science. International Conference Proceedings.*

Horváth, Gy. (2000). *Európai regionális politika.* Dialóg Campus Kiadó, Budapest-Pécs.

Horváth, Gy. (2015). *Spaces and Places in Central and Eastern Europe. Historical Trends and Perspectives.* London, Egyesült Királyság, Routledge.

Illés, I. (2008). *Regionális Gazdaságtan. Területfejlesztés.* Typotex Kiadó, Budapest.

Kőszegfalvi, Gy. (2008). Fiatal barátaink – Hét alföldi kisváros. *Területi Statisztika*, 3, 299–308.

Kovács, T. (1980). Magyarország településhálózata és fejlődésének főbb tendenciái. *Statisztikai Szemle*, 11, 1061–1080.

Kovacsicsné Nagy, K. (2010). Városiatlan városok sokasága. http.//mta.hu/tudomány_hirei.

Máthé, G. (1996). A polgári államszervezet kiépülése Magyarországon. *Rubicon*, 1–2.

MNL Somogy Megyei Levéltár V. Megyei Városok és Községek, 276. Nagybajom Nagyközség iratai (1946–1949).

Molnár, B. (2006). A magyarországi kis- és középvárosok globalizációs típusai. *Tér és Társadalom*, 4, 67–83.

Pirisi, G., & Trócsányi, A. (2006). Átalakuló kisvárosok a Dél-Dunántúlon. *Földrajzi Értesítő*, 1–2, 87–108.

Rechnitzer, J. (2008). Modellek a településhálózat lehetséges fejlesztési irányaira. *Falu Város Régió*, 3, 43–50.

SMÖH = Somogy Megyei Önkormányzati Hivatal. (2014). Somogy megye Területfejlesztési programja 2014–2020.

Szebényi, A., & Nagyváradi, L. (2009). Új város a pécsi agglomerációban – A kozármislenyi példa. *Területi Statisztika*, 2, 148–152.

Szigeti, E. (2002). *Község, város, jogállás. A magyar településhálózat közigazgatási térszerkezetének néhány kérdése.* Magyar Közigazgatási Intézet, Budapest.

T. Mérey, K. (1989). *A mai Dunántúli kisvárosok a dualizmus idején.* Manuscript. Pécs.

Tóth, J. (2008). Meditáció a városról és a várossá nyilvánítás hazai gyakorlatáról. Vitairat. *Területi Statisztika*, 3, 237–245.

Vaishar, A. (2004). Small towns – an important part of the Moravian settlement system. *Dela*, 21, 309–318.

Webpage of OECD urbanisation data. https.//data.oecd.org/.

Webpage of the United Nations. https.//unstats.un.org/unsd/databases.htm.

Zuzanska-Zysko, E. (2005). Economic transformation of small Silesian towns in the years 1990–1999. *Geographia Polonica*, 78, 137–150.

Section II

Initiatives

Tourism and territorial development: Dynamization strategies for an agrarian cultural landscape Guadalhorce Valley in Malaga

Lourdes Royo Naranjo

Introduction

Cascareros are one of the main signs of identity in the Guadalhorce Valley (Malaga) and constitute an architectural typology unique to this area of Malaga. However, the overall ignorance and the current state of conservation of these buildings have resulted not only in an architectural loss but also in the loss of values associated with the territory to which they are linked. The Cascareros owe their name to the main function they performed, drying the peels of citrus fruits, mainly sour orange. Cascareros were also used to dry other types of fruits characteristic of the area, such as almonds or figs. Their patrimonial recognition and consequent protection were detected at an administrative level recently, as some examples were included in the Pizarra PGOU protection catalog, written in 2011. Worst of all, many of these buildings had been converted into hotels or private houses. The base structures of the building increased their built surface considerably and they were adapted for residential use, forgetting what in its day gave rise to such presence in the territory of the Guadalhorce Valley.

The main objective of this work is the detection of a strategic line of development from these constructed elements scattered in the territory of the Guadalhorce Valley that occupied a position linked to the land and the orchards, becoming configurators of a unique Cultural Heritage. These arguments are part of a heritage reading and a development strategy, nowadays revisited by the scope of its management. The institutional efforts that are being carried out to revitalize the heritage of traditional activities have conferred a value of identity on the Andalusian territory, and also defined a strategic line by highlighting the tourist potential and pioneering initiatives of supra-regional cooperation, and by generating new tourist destinations for the rural environment. Consequently, we detect new models of public diffusion of the cultural and natural resources of the Guadalhorce Valley in a context where the crisis through which the theories of museology have passed bring experiences of renewal and new forms of management and dynamization of the heritage, and where the diffusion acquires a new role by which we could define it 'as a cultural management mediator between the heritage and society; therefore diffusion itself is not the information'.

Earth and water. The value of the orchard in the Guadalhorce Valley

The region of the Guadalhorce Valley is located in the centre-south of the province of Malaga (Spain) representing at a territorial level a bridge between the inner part of the province and its coastal area (Costa del Sol), which makes its geographical location a privileged value. Named after the Guadalhorce river, it is configured as an axis serving at the same time as the path and corridor of a cultural landscape where the orchards are scattered with farmhouses that conform the bottom of the valley and represent a living landscape. Among the mountains stand the villages of Alhaurín el Grande, Almogía, Álora, Cártama, Coín, Pizarra and Valle de Abdalajís, constituting the Guadalhorce region, a strategic point of the Costa del Sol, where the geographical references are the 'Sierra de las Nieves', the gorge of 'Los Gaitanes' and the set of reservoirs of 'El Chorro', as well as the mountains 'Montes de Malaga'.

The good climatic conditions and the abundance of water, as well as the proximity to the city of Malaga, contributed historically to the development and agricultural enrichment of the region of the Guadalhorce Valley (Fig. 1).

The Guadalhorce river, the longest and fastest-flowing in the province, created a space for the first settlers and its connection with the Mediterranean Sea represented a communication channel in contact with the diverse cultures that settled in our lands: Phoenicians, Greeks, Romans, Visigoths and Arabs who, above all the last-named, left their mark and exercised a decisive influence on agriculture – mainly irrigation agriculture. A great technological level in hydraulics was achieved, mainly in fields such catchment systems, water channeling and storage, as well as devices and mechanisms recognizable in the built heritage of the territory that we describe, notable in pieces such as irrigation canals, waterwheels and reservoirs.

Fig. 1. Guadalhorce Valley`s location in Malaga province. Source: Own ellaboration.

The citrus culture in the Guadalhorce Valley

Historically, the Guadalhorce Valley has also been known as the 'Valley of the Azahar' for its citrus orchards defined along the entire fertile plain of the river and its tributaries. The cultivation of citrus fruits, known in the valley since the Middle Ages, acquired enormous importance in the region during the contemporary period. Undoubtedly, without the boost given to agriculture during this period, it would be impossible to imagine the large expanses of orange and lemon trees in municipalities such as Álora, Pizarra, Cártama, Alhaurín el Grande or Coín (Fig. 2).

Currently, agriculture has undergone great changes influenced by the important tourist development of the Costa del Sol and the marked smallholding of farms, which leads to the progressive loss of importance of the agricultural sector[1]. Thus, most of the farmers diverted their main activity to other sectors (mainly to the services sector), transforming agriculture in many cases into a part-time activity.

In terms of heritage, the characteristics of the agriculture of the Guadalhorce Valley confer an added value not only on its products but also on its consideration as a territorial unit because it is one of the three areas of Andalusia, together with the Vega de Granada and the Alpujarras, being declared by the Junta de Andalucía as areas of interest for the development of organic farming[2].

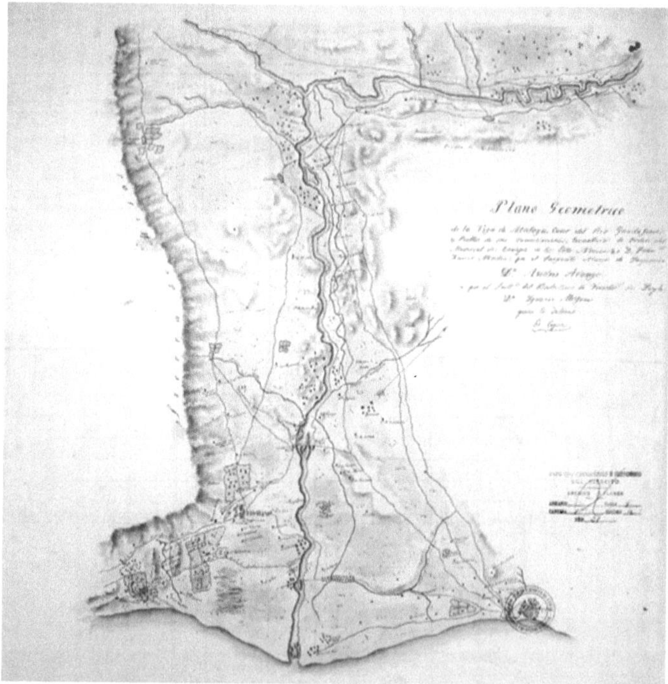

Fig. 2. Geometric plan of Malaga`s meadows, c. 1.826. Source: Historical maps archive of the Spanish Army Geographic Service.

1 Galacho and Larrubia 2003:159–190.

2 II Plan Andaluz de la Agricultura Ecológica, horizonte 2013.

Territory, landscape and traditional architecture. Values to work with

The orchard of the Guadalhorce Valley constitutes a territory whose fundamental values are recognized from the environmental, cultural, natural and landscape interest, and where work in the vegetable gardens and its relationship with traditional architecture is also highlighted. Following the letter of the reflection that is included in the *Plan Nacional de Arquitectura Vernácula*, Cascarero`s architecture is an architecture that has implicitly been part of the heritage for many years, associating itself with other aspects of which it is a part, as it is a part of the recognition of geographical diversity and landscapes[3]; while in other cases it has been identified as 'modest works that have acquired a cultural meaning over time'[4].

Since then up until today there have been numerous documents, both international and regional, that outline the value, usefulness and symbolic significance of this traditional architecture.

It is the architecture linked to the territory and the culture of the place into which it is inserted that confers a patrimonial recognition of what traditional architecture has meant throughout history. Its valuation has gone from being treated as a 'picturesque perspective'[5] to the 'expression of the cultural diversity of the world'[6].

Progressively, in the second half of the twentieth century, and as the concept and contents of Cultural Heritage expand, traditional architecture, in the context of the growing appreciation of Ethnological Heritage that reflects social practices and everyday (traditional) ways of life, has acquired a prominent role, partly as a consequence of the increasing risk of disappearance to which it is subjected by modern life: both by the loss of functionality to which we have alluded, and, above all, by its socio-economic aspect and loss of symbolic value.

---

3 Vv.Aa. *Recomendación relativa a la protección de la belleza y el carácter de los lugares y paisajes*, 1962.

4 Venice Charter 1964.

5 Athens Charter 1931.

6 As seen in ICOMOS, *Carta del patrimonio vernáculo construido,* 1999. Confirmed by the 12th General Convention in México in October 1999.

We should note in this regard an early recommendation of UNESCO which noted that 'contemporary civilization and its evolution lie on the cultural tradition of the people' and raised alarm about the disregard of this architectural heritage seriously affected by unbridled urbanism. It carries the explicit title of *Recommendation on the conservation of cultural property that the execution of public and private works may endanger*, Paris, 1968. Through the subsequent *Recommendation on the safeguarding of traditional and popular culture*[7], approved by UNESCO at its meeting in Paris in 1989, popular architecture will be considered one of the 'creations emanating from a cultural community based on tradition, expressed by a group or by individuals, that admittedly respond to the community's expectations as an expression of their cultural and social identity'.

The appearance of unique architecture in the Guadalhorce Valley: The cascareros

The *cascareros* or orange-peel driers are unique agrarian architecture of the Guadalhorce Valley in the contemporary age, because, although there are dryers throughout Andalusia, this typology and the characteristics with which they arise only occur in this region. Their particularity is possibly due to their high degree of adaptation or adequacy for the specific drying of the sour orange peel,[8] although other plants and fruits such as tobacco, figs and almonds were also dried. This type of construction had been known since the beginning of the nineteenth century, although it was in the second half of said century and the first half of the twentieth when it reached a greater boom, almost certainly due to the installation of the Malaga-Álora railway line in 1863.

The transport of this raw material played an important role, since it determined the strategic position of the driers, concentrated in points very close to the railway stations in areas where citrus crops were growing. The orange peel was transformed and exported abroad, mainly to

7 For further information check Núñez 2015:117.

8 The sour orange peel was in most demand due to the thickness of its skin and its aroma, but the fruit itself, due to its bitter flavor, was only used in some traditional soup recipes. The peels of other varieties of oranges were also dried and marketed, and it was common for country people to have their farms full of wires with drying peels waiting for the *aloreño* or *pizarrero cascarero* to arrive.

France and England, having various uses such as the production of gunpowder, essences, medicinal products, etc.

Fig. 3. *Cascarero* in the Guadalhorce Valley, 2018. Source: The author.

At the same time, the *cascareros* of the Guadalhorce Valley are located in spaces generally linked to farmhouses, as annexes or as independent buildings, although also next to places of work in which case they are usually smaller. Buildings may vary depending on the property, but always follow a common pattern that determines their type and function (Fig. 3). Structurally, these are simple, open, rectangular-plan structures erected as two or three floors supported by pillars whose number is related to the dimensions of the building. The ground floor is the only one partially closed with walls built of masonry, devoted to stables, storage or other livestock or laundry needs, while the remaining floors, which were accessed either by a staircase or a ladder, were intended to the drying of orange peel. The drying area does not have a not very solid frame, being formed by wooden beams and canes. The roof is made of canes on which Arab tiles are placed and featured one or two pitches with curved or flat tiles in the most modern buildings, while the void space between the ceiling and the canes was used as a dovecote. Up to two hundred people, mainly women *encascarando* (peeling oranges), and *verdeadoras* or *faeneras* (women rolling and packing oranges and lemons) could work in one of these *cascareros*.

Before the establishment of the Malaga-Álora railway line in 1863, the women of the Guadalhorce Valley were already leaving for work or *faena* (hence the name *faeneras*), where they were responsible for the collection of citrus fruits, almonds, olives and fruits as economic aid that contributed to the sustenance of the family. With progress, other tasks were added to these, such as the selection of products and packaging. We must emphasize with this the social value of the identity contributed by the work formerly carried out by the women of the Guadalhorce Valley in the *cascareros*. They handled the fruits of the field such as citrus, almonds, figs, etc. They were responsible for cleaning, wrapping and packaging the fruit later to be exported to Europe and other places.

Having registered this type of unique traditional constructions in the Guadalhorce Valley[9], twenty-eight *cascareros* have been inventoried, basically established in the fertile plains of the municipalities of Álora and Pizarra, as well as Cártama and Coín, but with little incidence, especially in this last. All *cascareros* documented in the town of Cártama are in a precarious state of conservation, if not in ruins, linked to workhouses, and used as tobacco dryers[10].

As stated in the informative report of the *Building and Archaeological Heritage Catalogue of the General Plan of Pizarra*[11] in its definition, purpose and field of application, this catalog aims to protect architecture throughout the historical framework, as well as archaeological remains and covers the entire municipality of Pizarra. Thus, this *Catalog* is configured as a critical tool for the protection of this heritage as a historical memory of the urban and architectural culture of the town. Its purpose is to guarantee adequate protection for all elements of the architectural, urban and archaeological heritage of a historical and cultural nature that contributed to the formation of Pizarra's identity and its spatial, urban and architectural quality.

---

9 Vv.Aa., *Documento único del Plan General de Ordenación Urbanística de Pizarra. Catálogo de patrimonio histórico artístico, edificado y arqueológico*, 2011.

10 As is the case of Cortijo Supervieral or Cortijo-venta Juan Sánchez.

11 Single document of *Plan General de Ordenación Urbanística*, municipality of Pizarra as per agreement of Provincial Commission for Territorial Planning and Urbanism, 30th March 2011.

Tourism and heritage. Territorial dynamization strategies and social participation

The first attempts to disseminate the natural and cultural heritage of the Guadalhorce Valley are linked to the current role of the *Valle del Guadalhorce Rural Development Group* (GDR), which focuses on the attempt to recover and enhance traditional architectural heritage such as the *cascareros*, based on a series of value-added strategies to raise awareness of their importance at the level of heritage identity. As stated on its website[12], the GDR is a non-profit organization that, although finding part of its *raison d'être* in the management and execution of rural development aid programs in the Guadalhorce Valley (*Proder I* from 1998 to 2001; *Leader* and *Proder Andaluz* between 2002 and 2006 and *Feder* until 2015), has its true essence and priority objectives in the promotion of and commitment to sustainable development.

Economic, social and cultural revitalization and environmental awareness, as well as cooperation, are the main axes of the work of an entity that brings together the identity of the Guadalhorce Region and that has, after fifteen years of work, become a benchmark at a national level. But perhaps the most important thing is to highlight the method of work, which is based on citizen participation since it is considered that only by knowing the opinion of the inhabitants of the territory can rural development be carried out in a sustainable way. For this, the work is structured in committees focused on the following aspects: Agriculture and Livestock; Training and Employment; Industry and Commerce; Heritage; and New Technologies. The activity of these commissions has resulted, thanks to the GDR Valle del Guadalhorce, in the constitution of a series of associations and forums aimed at studying the needs of different economic, social and cultural sectors and, once these needs are known, establishing lines of action destined for the improvement, dynamization and structuring of the social, economic and cultural life of the Valley.

Although an advance in the appreciation and interpretation of cultural heritage as a tourism resource is perceived, there are still serious difficulties in working with a not-so-centric heritage, either in the

---

12 http://www.valledelguadalhorce.com/seccion_asociacion.php. (latest search 23/07/2018).

countryside or territorially distributed, but which, if interconnected, can represent one of the greatest opportunities that tourism could offer for the development, enhancement and regeneration of these territories and their heritage identity.

The lines of work initiated from the *Strategic Planning Document of the Guadalhorce Valley* are linked to the *General Sustainable Tourism Plan of Andalusia*, which defines the following challenges for this area of Malaga:

- Pace the growth rhythm of the capacity of reception of the tourist space with that of overnight stays and improve their level of use.
- Maintain and reinforce the tourist attraction of our destinations.
- Differentiate Andalusian destinations and products to unlink their competitiveness from the downward price race.
- Maintain and professionalize stable and quality employment.
- Reduce the fragmentation of the identification of territorial brands and the productive structure.

The link between natural tourism and active tourism occupies almost all of the publications and official tourism promotions that reflect the heritage values of the landscape of the Guadalhorce Valley[13]. However, at the level of management and enhancement of the said heritage, we detect an offer lacking knowledge of the natural and cultural heritage in its territorial, landscape and specifically cultural components. At a pedagogical level, the efforts are related to some publications on issues of a traditional nature. Some tourism publications that promote the cultural values of the Guadalhorce Valley (Fig. 4) very briefly reflect what is happening in terms of heritage and tourism in the province of Malaga at a general level. Malaga is positioned as a capital of international tourism and moves within parameters where tourism and touristification go hand in hand. However, the problem of mass tourism in historic centers is not a recent issue at all.

---

[13] Galacho and Luque 2000: 25–58.

Fig. 4. Tourism leaflet 'Territorio Caminito del Rey, in Guadalhorce Valley, Malaga'. Source: Junta de Andalucía.

Valle del Guadalhorce Agricultural Park: First steps for the management of a heritage to be discovered

Theoretically, we must understand that both the dissemination and the management of the heritage involve a complex process that encompasses documenting, assessing, interpreting, manipulating, producing and disseminating not only the object itself but a comprehensible and assimilable model, in its relationship with its historical past and its present environment to be part of history and therefore of identity. Among the most recent innovations in business matters, we detected an initiative that strengthens the value of the territorial unit in its dynamic role, through the Agrarian Park Valle del Guadalhorce.

This initiative aims to develop a tourist destination that offers leisure complementary to the sun and beach of the coast of Malaga, this time more related to countryside tourism or rural tourism, in a geographical setting characterized by its rural and agricultural nature, but also where patrimonial issues as obvious as those mentioned earlier are of special interest. As defined from the Agrarian Park Valley of the Guadalhorce you can enjoy endless activities, from walking through a farm of olive trees or enjoying the flowering of the orange trees, to visiting cheese factories or wineries, visiting the historical monuments of the villages or tasting the traditional cuisine of the Algarbía, based mainly on fresh products from vegetable gardens[14]. The creation of the Agrarian Park Valle del Guadalhorce has, among its objectives, the maintenance of this identity against speculation and urban and anthropic pressures, while at the same time responding to a demand of urban origin that requests open quality spaces nearby. As for its structure, the Guadalhorce Valley Agrarian Park is formed of seven villages with an area of 772.8 km^2 and a population of approximately 99,000 inhabitants, distributed in the municipalities of Alhaurín el Grande, Cártama, Coín, Álora, Pizarra, Almogía and Valle de Abdalajís.

The objectives of the recognition of the Guadalhorce Valley as an Agrarian Park are defined as follows:

- To strengthen the local economy by fostering employment and self-employment; to energize the socioeconomic activity of the Guadalhorce Valley by helping to create green jobs and strengthening agricultural activity and short marketing channels.
- A green corridor: This territory also has great importance as a connection space, since it serves as a link between different natural spaces, such as the Protected Natural Area of the mouth of the Guadalhorce river, with the high stretch of the river and the Biosphere Reserve Sierra de las Nieves; and between the different mountain systems that surround it.
- Conservation of biodiversity: The existing nature of the Guadalhorce Valley is a characteristic feature with a great diversity of ecosystems in which unique species of both animals and plants can be found. In addition, the Agrarian Park Guadalhorce Valley contributes to the conservation of agricultural biodiversity by maintaining local crop varieties that are better adapted to the environment.
- Landscape generator: The landscape in the Guadalhorce Valley is very diverse; mountains, areas of both rain-fed land and irrigated crops, and large areas of water, such as the Guadalhorce River and its tributaries (Fig. 5).

---

[14] search (20/07/2018).

Fig. 5. Landscape at the Guadalhorce Valley. Source: The author.

Education and heritage. A path for dissemination

The last link in this chain is the work we present for the dissemination of the heritage as a factor of development that must be inscribed in a broad reflection on ways of revitalizing the space to promote a more balanced distribution capable of preserving territories with great heritage wealth, both historical and natural. Without a cultural assessment of the territory capable of highlighting the uniqueness of its human resources and stimulating the confidence and creativity of the community, it is difficult for our community to take off in its economic development[15].

The dissemination of heritage (natural or cultural) does not consist only of 'exposing' the visitor to some objects with panels, posters, audiovisual media, etc. but of transferring to him or her, regardless of age or cultural level, a vision as close to reality as possible[16]. The dissemination

15 Ruiz Parrondo 2010: 165–177.

16 Martín 1996.

of the heritage means allowing the public to receive an objective message about what they see, providing them with a knowledge of their past, showing the topics through proper communication, and trying to generate a desire to repeat.

The plotline of our proposal for the diffusion of the Valle del Guadalhorce Agrarian Park defends the enhancement of the natural heritage, its landscape, as well as the cultural and architectural identity of the orchard in the Guadalhorce Valley, in order to design the dissemination of these values, departing from a series of activities oriented to different educational levels. From this orientation, the Agrarian Park Valle del Guadalhorce would be enhanced for the population by the knowledge and heritage value of all the cultural heritage of the territory, as well as the territorial identity of the region.

With our initiative and from education about the heritage, we try to manifest and highlight each and every one of the aspects we have recognized in the Guadalhorce Valley, where the orchard is the center from which we start to intensify with greater strength each of the representative aspects of a unique Cultural Heritage. Our territorial dynamization proposal, starting from the figure of the vegetable garden, defines the knowledge[17] of an architectural element particular to this land such as the *cascarero*, traditional architecture that links a series of activities, starting with work in workshops accompanied by a dissemination of the knowledge acquired around the natural and cultural heritage of the territory. In this way, we transform the land management model in a sustainable way in the economic, social, environmental and patrimonial spheres, by using the agro-ecological and cultural resources[18] of the territory with the involvement of the different actors and citizens.

Currently, the first efforts for the recovery of the orchard as the backbone of the Guadalhorce Valley are being carried out by the GDR management, and our project responds to the very latent need to raise awareness of the value of identity of the territory. We intend to approach other positive and well-managed forms of heritage interpretation in order to spread the cultural values associated with the territory in a shared effort, where the vision to be transferred has to do with a larger, more

---

17 Tilden 2006.

18 Morales 2001.

territorial and complex scale which makes possible a knowledge of the values of the land and its landscape.

Bibliography

Galacho, B., & Larrubia, R. (2003). El sector agroalimentario malagueño ante el proceso de globalización oportunidades, deficiencias y retos. *Baética: Estudios de arte, geografía e historia*, 25, 159–190.

Martín, M. (1996). Reflexiones en torno a la Difusión del Patrimonio Histórico. In *Difusión del Patrimonio Histórico* (pp. 13 – 27). Instituto Andaluz del Patrimonio Histórico.

Morales, J. (2001). *Guía para la interpretación del Patrimonio.* TRAGSA.

Ruiz, A. (2010). Interpretación y difusión: dos formas diferentes de ver el patrimonio. *Arqueología y Territorio*, 7, 165–177.

Tilden, F. (2006). *La interpretación de nuestro patrimonio.* Asociación para la Interpretación del Patrimonio.

Vv.Aa., *Recomendación relativa a la protección de la belleza y el carácter de los lugares y paisajes*, 1962.

Planning and legislation

Plan Nacional sobre Arquitectura tradicional. Ministerio de Educación, Cultura y Deporte. Gobierno de España. http://ipce.mcu.es/pdfs/PNArquitecturaTradicional.pdf.

Atenas Charter, 1931.

Venice Charter, 1964.

Carta del patrimonio vernáculo construido, 1999. ICOMOS.

Plan General de Turismo Sostenible de Andalucía 2014–2020. Orden de 23 de julio de 2014, por la que se somete a información pública la propuesta del Plan General del Turismo Sostenible de Andalucía 2014–2020, y el Informe de Sostenibilidad Ambiental.

II Plan Andaluz de la Agricultura Ecológica, horizonte 2013.

Documento Estratégico de Planificación del Valle del Guadalhorce http://www.fundacionmadeca.es/images/pdf/Mesas_Territoriales/Integracion_NERA_Valle_del_Guadalhorce.pdf.

Documento único del Plan General de Ordenación Urbanística del municipio de Pizarra conforme acuerdo de la comisión provincial de ordenación del territorio y urbanismo del 30 de marzo de 2011. http://pgou.pizarra.es/DU/home.html.

Websites

http://www.pizarra.es/

http://www.valledelguadalhorce.com/

http://www.valledelguadalhorce.com/huerta/seccion-nuestra huerta-Agricultura

http://www.parqueagrarioguadalhorce.com/visitanos/actividades/

http://guadalhorceturismo.com/

http://juanito-orange.wix.com/paginaoficial#!contact/c16fm

Viterbo. Heritage and future

Manuela Raitano

co-author Paolo Marcoaldi

Introduction

In January 2016 the Municipality of Viterbo decided to commission a research concerning the historic centre of Viterbo from the Department of Architecture and Design of Sapienza, University of Rome. The design group, coordinated by Orazio Carpenzano, was called to elaborate a "guidance document to identify principles and instruments of procedural and technical nature for the redevelopment, enhancement and preservation of the historic centre of Viterbo" (Fig. 1).

Fig. 1. Aerial view of the city of Viterbo.

The initial intentions of the Municipality were essentially to develop in a short time (January to November) a study capable of proposing strategies for the future of the most interesting heritage sites of the town.

The work started without specific guidelines, but with the ambition to plan safeguarding actions for an impressive historic-artistic heritage[1] that was never really planned before: in fact, the previous "Plan for the historic centre" (drawn up in 1988) was unfortunately never adopted, and therefore there had been no other plans at all. Until that moment, local administrations had proceeded essentially following two different kinds of plan, mutually incompatible: 1) a variant of the "General Plan" of 1979, which considered the center as a whole, entirely subjected to a strict protection regime; 2) extraordinary programs that, contravening the "General Plan", used to be started when public funding became available.

Methodology

As academic team, we decided to intervene in a context so uncertain and fragile using a consolidated methodology, on the basis of previous experiences in the planning of historic centres, starting with the urban plan of Città di Castello[2].

As a first step, the group involved other local expertise, establishing a scientific committee entirely composed of prominent figures rooted in the territory: university professors, artists and professionals who dealt with the subject "Viterbo" according to their specific knowledge.

The members of this scientific committee immediately participated in the formation of the research program, becoming over time an indispensable link with the city, capable of rooting the design team in the specific site of the project (Figs. 2 and 3).

---

1 Pinzi 1916, Scriattoli 1920.

2 Carpenzano 1993.

Fig. 2. The medieval city walls (photo by Francesco Galli).

Fig. 3. "The lodge of blessings" in the papal Palace of Viterbo (photo by Francesco Galli).

Once the main members of the research group had been defined, we moved on to the analytical interpretative study of local urban events. To tackle such a complex urban phenomenon, it was essential to place it in a wider context, comparing it with other models that had at least a sufficient number of comparable characteristics. The analogical criteria used to select urban settlements comparable with Viterbo were essentially three: first, towns similar in the number of inhabitants were chosen; second, towns within a common territorial context, the Central Apennines; third, towns with a similar medieval matrix of the main urban system. Based on these three criteria, the cities identified for comparative studies were: Arezzo, Ascoli Piceno, Orvieto and Siena.

Ever since the preliminary morphological analysis, it was evident that Viterbo represented a singular case-study. In fact, the historic centre of Viterbo is not situated in a raised position or in a strategically dominant place, as in the case of the other sample cities. It is because the main axis of *extra-moenia* expansion developed along the slopes of the *Cimini* Mountains: for this reason, the modern neighborhoods (denominated *Cappuccini* and *Murialdo*) are now positioned at a higher level than the historic city. This orographic inversion in the relationship between the center and its periphery is made even more dramatic by the lack of a critical distance between the inside and the outside of the city. In fact, suburban expansion grew in contact with the center, completely canceling the spatial and figurative qualities of the medieval city walls, a gigantic stone threshold conceived to be a limit and a passage at the same time (Fig. 4).

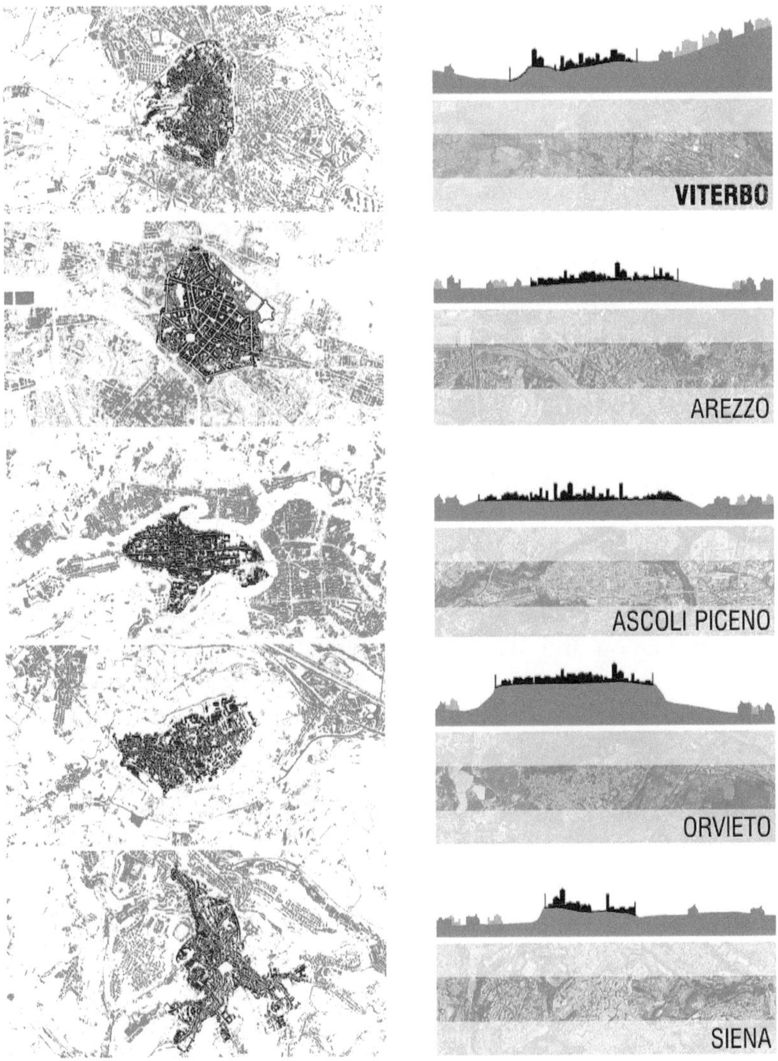

Fig. 4. Comparative analysis between Viterbo and the sample cities.

These facts determine an equivocal visual relationship, since the walls no longer have the power to separate the inside perceptibly from the outside. The historic centre disappears from the periphery and, in turn, it is ambiguously observed from the outside. Otherwise, the low number of residents within the historic city of Viterbo, combined with the relatively low average age of the resident population, determine an ideal socio-cultural context to experiment with unconventional housing policies, verifying the possibility of introducing, in the historic medieval fabric, new innovative typological and organizational forms.

A further important contribution, decisive for consolidating the theoretical assumptions of this collective work, came from the local community: individual citizens and sector categories were involved in drafting the Masterplan through a multitude of initiatives, preliminary meetings, surveys and public exhibitions.

However, the participatory vision of the urban project had a fundamental but limited role: the local community helped us in verifying the instances of the research, but they did not have a role of guidance, since this role was implicit in the work of our Scientific Committee.

Masterplan assumptions

The Masterplan for the historic city centre of Viterbo is a strategic document aimed at indicating the main directions of development for the enhancement of the historic centre, in order to guide the policies of transformation and active protection of the city. The term "strategic" represents the peculiarity of this predictive tool, agile and versatile, which looks more like an "instruction manual" than a real urban plan. It is a type of tool that does not descend, in genealogical line, from the typical Italian urban plans for historic city centres, since it represents a radical revolution in our traditional planning practices.

Simplifying, we can divide the plans for the historic centres into two large families: the first inspired by the Bologna plan by Pier Luigi Cervellati (1969); the second dating back to the Pesaro plan by Carlo Aymonino and Raffaele Panella (1970/74). These two different plan visions corresponded, as is well known, to two different ideas of protection and two different conceptions of the role of historic centres in the body of a modern city.

In Bologna, the theme of safeguarding historic fabric was strongly affirmed, with the great merit of making the planner's goal very clear, as the historic city was presented as a physical body that had to be preserved as a whole. In Cervellati's vision, substitutions or urban infills were admitted with difficulty, while only specific actions of "conservative rehabilitation" were admitted; the latter, as a sort of urban *anastylosis*, were used to compensate for the integrity of the individual urban part, thus contributing to the so-called "modernization" of a city centre. Nowadays, we know the cultural consequence of this approach: after the Bologna plan, the historic centres, according to many, became monuments in themselves. This fact, if on the one hand contributed to the growth of a strong culture of heritage protection, on the other hand contributed to freezing the natural development of a city as an organism: if in fact the historic city is an artefact to be kept *in toto*, then it must be restored as a whole, as a single compact monument, not designed as a contemporary living organism.

In Pesaro, on the contrary, Aymonino and Panella designed a plan starting from the scale of architecture. The normative aspects of the plan were balanced by the figurative ones that supported some targeted but significant architectural interventions that reshaped the historic centre (and above all its margins) giving it a new form. According to the vision of the designers, the "new" not only had the legitimacy of being in the "old", but was also identified as the only possible way to rewrite and to continue to keep the ancient alive. Which, in turn, had to be read as an overlap of two layers, the fabric and the emergencies, both configuring the shape of a city as a dialectic of background and figure, like Aldo Rossi stated in its famous book *Architecture of the city*.

This second position has the merit of overcoming the mere bi-dimensionality of the plan; nevertheless, the limits of this approach must also be highlighted. They can be summarized in a single question: can the new architectural pieces that redesign the historic city be governed by the plan? In other terms, "forcing" the urban design scale by introducing elements of architectural scale, does not risk provoking further rigidity that compromises the architectural quality to come? This risk was avoided, in the specific case of Pesaro, thanks to the fact that the two planners were also the architects involved in the design of the architectural parts that they planned; but in how many cases did we see plans that prefigured incongruous shapes, strange architectural forms, established by technicians longing to deliver to history their personal "expression"?

Therefore, in the attempt to experiment a third way, new instruments must be explored. We call them now "Masterplans", that means in brief "pilot projects". These are documents whose normative content is reduced to the bone, but which prefigure the possible intervention strategies trying to define their interactions and their thematic overlaps. From a technical point of view, the Masterplan we propose for the city of Viterbo does not design the places, but proposes design actions that will make possible the transformation of the places; it does not establish restrictions or prescriptions but indicates the range of possibilities for the interventions; it is not a plan of negations, rather it is a schematic representation of the total amount of the opportunities. So, we will say that the masterplan for the historic centre of Viterbo is an "action plan".

The "action plan" is a document that does not draw the "form" of the historic city on the level of conservation of its historical physical matter, nor on the level of the introduction of new architectures. Presenting itself as a set of strategies, it solves the dichotomy expressed by the two famous cases of Bologna and Pesaro. By virtue of this, it can remain legitimately bi-dimensional without thereby mortifying the spatial vision of the city, since its strategic definition allows one to get very quickly to the phase of architectural intervention by activating project financing procedures, or by funding international design competitions to be carried out in an open or restricted form, depending on the complexity of the themes. It is also important to point out that the approval of a strategic plan is a useful pawn to increase the possibilities of a Municipality to get European funding, to start singular interventions according to the plan.

In conclusion, a Masterplan is a flexible tool, not articulated at a normative level, but studied in depth on a thematic level. This is the masterplan for the historic centre of Viterbo in its formal structure.

Theoretical frame

The proposals drawn up by the Masterplan are based on three basic choices that are indispensable for us, which have guided the entire work: 1) the strategic importance of the city walls in the redevelopment policies of the city; 2) the enhancement of the natural landscape; 3) the idea of a renewable city, obtained throughout calibrated infills of contemporary new architecture in the body of the ancient city.

So, the theoretical framework of the Masterplan stands between these two terms: "heritage" and "future". For us, these are the two poles of a delicate balance between the instances of "protection" and the necessity of "transformation"[3].

Viterbo, like many other Italian historic centres, represents an urban heritage not qualified as it would deserve. Viterbo has a unique medieval historic centre[4], in the shadow of the proximity of Rome. The capital, in fact, represents a huge tourist attractor that steals from Viterbo the "brand" of a city whose identity is also tied to the history of the popes. The work we have done tries to reverse this situation, using the proximity of Rome as an opportunity for developing a high-quality tourist offer, as well as new scenarios for everyday contemporary life in the city.

The theoretical framework of the entire work is based on the combination of heritage and future. Special attention, as we said, was given to the theme of the enhancement of the important natural component of the city of Viterbo (the *Valle Faul*), in synergy with the network of the other historical-archaeological resources of the *Tuscia* territory[5]. Moreover, although it is a work that concerns the historic centre, special attention has also been paid to outside the perimeter of the city walls, to illuminate the relationship that the historic centre establishes with the entire city and with its territory. This has been made in the belief that a serious policy of revitalization of the historic city centre cannot ignore it as an urban part connected with its surroundings.

In this sense, the physical location of the *Valle Faul*, crossing the boundary of the city walls and entering the historic town, appears to us as a potential, extraordinary link between the consolidated urban landscape and the *extra-moenia* landscape that leads from the city towards the thermal baths and towards the ruins of Castel d'Asso.

---

3 De Vita 2012.

4 Bentivoglio and Valtieri 1982, Bonelli and Bonelli 2005.

5 Brancati 1934.

Contents

From the point of view of the contents, the document consists of thematic elaborations (residence, tourism, landscape, etc.), each of which supports the analytical reflections (graphic synthesis of the actual state) to the design intuitions (development lines and transformative scenarios).

As said before, the proposals elaborated by the Masterplan are based on three fundamental basic choices which have oriented the entire work: the strategic importance of the city walls in the redevelopment policies of the city; the enhancement of the dominant landscape within the walls; the idea of a renewable city through a controlled infill of new buildings, according to contemporary languages and functions (Fig. 5).

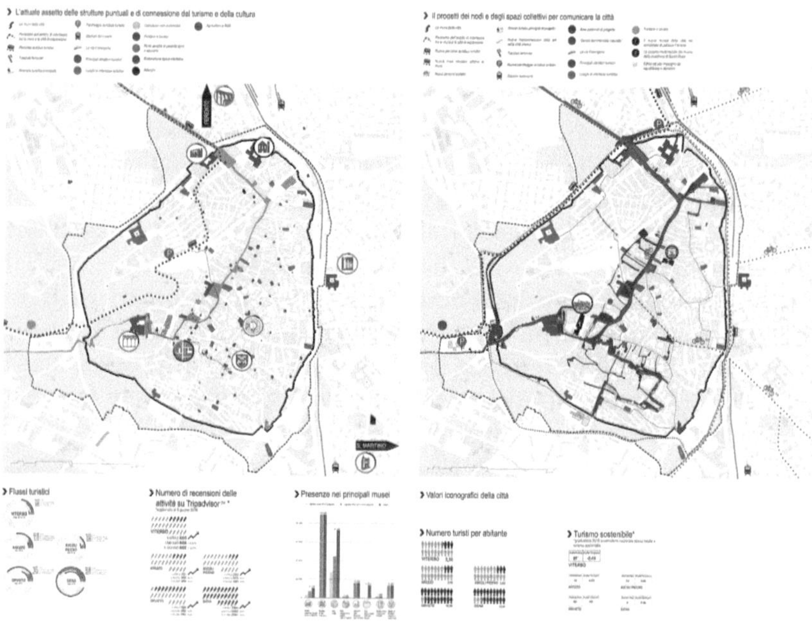

Fig. 5. Masterplan tables. Sample.

The territorial strategies

The first steps, as mentioned above, had a prevalent analytical nature: we did not introduce any new design, but we specified the potentialities implicit in the Viterbo area.

The observation of planning documents at the regional and provincial scale confirmed our intuitions, based on the considerations that Viterbo is a city with a strong landscape potential linked to thermalism, to the presence of the *Via Francigena* and to the surrounding Etruscan sites. To summarize, there is a triad of attractions linked to three different qualities of the landscape: the natural landscape, the historic landscape and the archaeological landscape.

The scale of the municipality

At the municipal scale, some actions can be implemented in a short time with small and limited investments. These are actions that all come from outside the perimeter of the city walls but, in our opinion, they are essential for laying the perspective of the development of the historic city centre on solid foundations.

Therefore, we planned to set up a network of cycle-pedestrian routes, linking the historic centre of Viterbo with the constellation of the minor historic centres that surround the city. These paths can be realized through small operations aimed at their implementation, to design a "ring" system of connections between the main connection routes, and between these and Viterbo. These connections are already almost all existing in the form of secondary paths, often unpaved and currently badly indicated. As the Masterplan's drawings illustrate, a few targeted actions would allow us to move from a radial distribution of the connections (where one has to go and come back along the same route) to an annular circulation, more diversified and better linked to a wider use of the territory.

The urban scale

The lack of a post-unitary urban plan and the unbalanced expansion along the south-east/northwest side of the city had the merit of leaving untouched the western sector of the city, outside the city walls[6]. It is an urban quadrant characterized by the predominant shape of the landscape

---

6 Valtieri 1977.

that, like an open hand, creeps into the historic centre through the *Valle Faul* to cut off the margins of the compact urban fabric.

The *Valle Faul* is now an underutilized resource, but it is also a magnificent opportunity, a keystone to approach the rethinking of the role of green in the historic centre as in the whole city. It can become, concretely, a physical place through which to realize the connections of the soft mobility intra and *extra-moenia*, a powerful hinge situated between the city and the territory.

This natural presence in the city centre is therefore very powerful in Viterbo and it should be considered as the main resource of the project, for the enhancement of the historic centre. All design strategies go in this direction, preserving the continuity of the main environmental systems.

Living "in the centre"

We also carried on a careful analysis of urban fabrics, making a census of the disposable voids in the ancient city blocks, searching for public buildings in a state of underutilization. Our idea of urban regeneration intends to encourage small and targeted operations of infill, involving also private investors, to reuse the disused public buildings. The new functional destinations will be targeted at students and on a particular group of users: the so-called "*grey area*" of the population, made up of those who do not have the requirements to get social housing, but do not have the income to rent or buy a house.

In this framework, the network of proximity services is strengthened by making available to new users a part of the public assets currently underutilized, facilitating the use of services in the historic centre and accentuating the urban vitality.

Recovery strategies for existing heritage

The masterplan of the historic centre moves from the consideration: no protection is really a solution if it is not accompanied by an idea of use (or better, reuse) of the building fabric of a historic artistic centre. A real interest in the past must necessarily involve a full responsibility towards the heritage, which for this reason must not only be cared for and safeguarded, but also reinvented, rewritten and reused; in a few words it must be re-designed.

According to these considerations, one of the masterplan's strategic works records the amount of public property currently underutilized or even abandoned, drawing a sort of "map of underutilisation". On the one hand this map represents the amount of useful resources currently dissipated but, on the other hand, it may represent a true opportunity for the transformation of the city centre, showing with evidence where it is possible to intervene.

Within this census of forgotten places yet available, new innovative collective functions are placed: public theaters, accommodation for students of the university campus of *Tuscia*, reading rooms related to the local University, new museums dedicated to the history of this ancient city, etc.

The proposed hypothesis of reuse alternates functions related to the residents with receptivity and tertiary functions, such as services for culture and leisure. Each of these interventions can be started independently, although they constitute all together a unitary vision that corresponds to an overall program of reuse of the historic centre.

The city walls project

As already anticipated, one of the main topics of the masterplan concerns a different perception of the medieval city walls, which are considered as the main place of interface and exchange. For this reason, it was necessary to expand the normative perimeter of the historic centre, to include parts of urban areas located outside the medieval walls, comprehending them within the walls project, in order to create a green ring of varying thickness that follows in parallel the perimeter of the walls (Fig. 6).

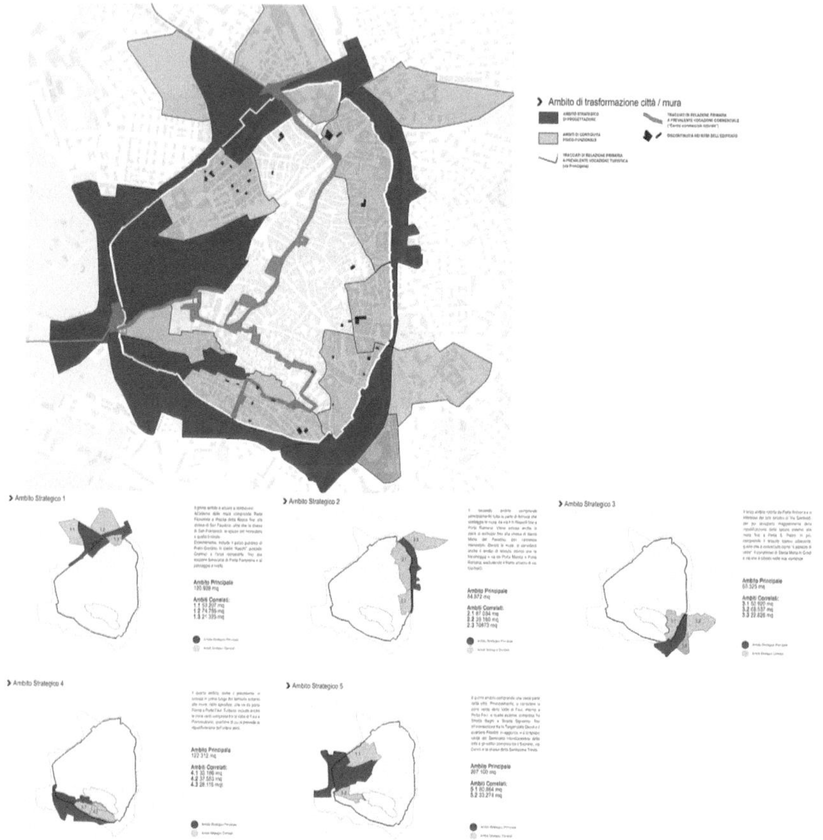

Fig. 6. Masterplan tables. The city walls project.

The city walls project also foresees, as clearly indicated in the tables related to the landscape, the recovery of the ancient walls and their openness to the public, as points for viewing the city and its territory from above.

The project incentives – as an optimal solution – the hypothesis of putting the railway section between the stations of *Porta Romana* and *Porta Fiorentina* completely underground, and the construction of a cycle path on the current site of the tracks.

In all the different areas of intervention the main actions to be undertaken will concern the improvement of accessibility to the historic centre and the strengthening of the connections and the nodes of exchange between the different urban systems, inside and outside the walls.

The reorganization of the driveway circulation is foreseen all around the walls, the establishment of a circular line of shuttle-buses and the creation of underground parking, immediately placed *extra-moenia*, but with pedestrian exits directly *intra-moenia*.

International design workshop

Parallel to the drafting of the plans, it was possible to verify the contents and the theoretical assumptions of the research through the scale of an architectural project.

An international architectural design workshop (held in Viterbo in October 2016) was directed by Orazio Carpenzano. The jury that evaluated the final designs was chaired by Herman Hertzberger.

During the workshop, a huge group of architects and Ph.D. candidates imagined possible scenarios on five specific areas previously identified by the research group as areas of interest. The five sites were: The *Papal Gardens*; the *Sacrario Square*; the *Fortezze*; the *Rocca Square* and the former *Palazzo Farnese* on the Cathedral hill.

The design themes concerned specific formulations indicated in the masterplan: at the *Palazzo Farnese* the possibility of converting an ancient abandoned hospital into a new Museum was verified; for the *Sacrario* square the main target was to mitigate the surface parking and integrate it with functions related to leisure and recreation; the *Papal Gardens* have been imagined as an area through which to realize pedestrian access to the Papal Palace, involving the path of the *Via Francigena* that exists in that area; underground parking has been planned in the immediate surroundings of the former church of *Santa Maria delle Fortezze*, outside the walls, while a strategy of urban densification was envisaged within the walls; lastly, there was an attempt to redesign the public spaces of the north-west sector of the historic city, working on a system of three squares that included, in addition to *Piazza della Rocca*, the spaces in front of the papal stables designed by Bramante, and the churchyard of the *San Francesco* church.

In all these proposals, new languages fit into the old ones without any formal or semantic ambiguity. Integration, if necessary, does not deny existing temporality, but adapts itself to the needs and expectations of contemporary living (Fig. 7).

Fig. 7. Around the acropolis; winning project of the international workshop (arch. F. Balducci and A. Farris).

The city, and with it its ancient centre, is a process in continuous becoming, in which the new is not only justified in staying alongside the ancient, but rather is the only possible way to rewrite and to continue to keep alive the past.

Conclusion

The results of this research are now published in a book titled *Per la città di Viterbo. Il masterplan del centro storico, direzione scientifica di Orazio Carpenzano*[7].

---

7 Marcoaldi 2018.

The Masterplan was also adopted by the municipality of Viterbo during one of the last council meetings of the administrators who governed the city in the period 2013–2018.

The real hope is that this research, which collects not so much a series of specific projects, but above all a new vision for the historic city of Viterbo, will be not only the synthesis of a beautiful participatory path, but the prelude to a new beginning.

The idea is to find a sustainable solution for tourism in the old city: tourism that must grow because it is an important economic income but, at the same time, tourism whose impact must be reduced in order to reduce that process of commodification of heritage that must be avoided.

The possible development scenarios of the Masterplan are then substantially two. The first step involves the formation of a permanent workshop on the historic centre of Viterbo ("Viterbo City Lab"), a multidisciplinary structure that, based on the contents of the plan, can control the quality and feasibility of the main initiatives concerning the historical city.

The second process of implementation of the Masterplan consists of a list of the main strategic interventions that the new municipal administration is required to complete in the coming years. This time schedule of the main public and private initiatives concerning the historic centre does not exhaust and does not fully capture the true legacy of the Masterplan.

As time goes by, it has been increasingly evident that the city has now lost interest in wide-ranging initiatives: the administrators changed after the elections and Viterbo citizens look with suspicion to everything that does not have the single purpose of solving the contingencies of the moment. The Masterplan has tried to take care of a very significant inheritance. But the destiny of Viterbo and its old town is inextricably linked to the sharing of a cultural vision that has considerably reduced since the mid-1970s, just when Pier Paolo Pasolini invoked for the whole region "a cultural turning point, a slow development of conscience"[8].

The research undertaken by our group transcends the boundaries of a traditional planning, and stands as a cultural model that measures urban problems within the conceptual frame of architecture, seeking to

8 *Cara Tuscia*, interview with Pier Paolo Pasolini, published in the newspaper "Il Messaggero" on 22 September 1974.

renew, throughout a plurality of knowledge and reading tools, the sense of everyday life inside the historic city.

The real hope is to give a concrete perspective to an inattentive city, which perhaps no longer looks to the future with the awareness of its important past, suffering for a lack of perspective in a present poor in ideas, poor in culture, poor in architecture.

References

Bentivoglio, E. and Valtieri, S. (1982) *Guida a Viterbo.* Dedalo.

Bonelli, L.P. and Bonelli, M.G. (2005) *Dal Castrum Viterbii alla Civitas Pontificum. Arte a architettura a Viterbo dall'XI al XIII secolo.* Stampa Agnesotti.

Brancati, V. (1934) *Singolare avventura di viaggio.* Mondadori.

Carpenzano, O. (1993) *La struttura urbana di Città di Castello. Formazione e trasformazione del centro storico.* Petruzzi editore.

De Vita, M. (2012) *Città storica e sostenibilità. Historic cities and sustainability.* Firenze University Press.

Marcoaldi, P. (2018) *Per la città di Viterbo. Il masterplan del centro storico, direzione scientifica di Orazio Carpenzano.* Quodlibet.

Pinzi, C. (1916) I *Principali Monumenti di Viterbo. Guida pel visitatore.* Stampa Agnesotti.

Scriattoli, A. (1920) *Viterbo nei suoi monumenti.* Capaccini.

Valtieri, S. (1977) *La genesi urbana di Viterbo.* Officina.

The white coal town, the emergence and decline of textile industry in Naoussa

Nikolaos Leonidakis

Introduction

Naoussa is a small town in central Macedonia, about 90 km west of Thessaloniki. Situated at the foot of Mount Vermio, and having a handicraft textile tradition[1], its entrepreneurs took advantage of the waterfalls of the Arapitsa River that runs through the town, in order to develop one of the most important industries in the European part of the Ottoman Empire.

The subject of this paper is the presentation of the emergence and decline of the textile industry in the town of Naoussa. Data obtained from the research of the master's thesis, «The financing of a 'bandit-stricken' industry with Marshall Plan funds, the case of Pennie wool spinning mills-Lanaras Bros SA», were used, as well as related local literature, newspapers articles and interviews.

The emergence of Naoussa's textile industry was based on a particular local social structure combined with the existence of a large domestic market. After the incorporation of the region into the Greek state, the distinctiveness of the town in relation to the smaller provincial pre-war industrial centres of Greece[2], was that despite the change of political and economic conditions in the twentieth century, and the total destruction of its industrial facilities during the Greek civil war, some local industries

[1] Stougiannakis 1924: 62–63, where it's reported that the production included sajak, woolen fabrics, silk-linen and towels. Also Vasdravellis 1967:281–295, on a list of the properties that were confiscated by the Ottoman revolution in October 1822, 4 tanneries, 3 paint shops, 19 workshops and 27 mills were recorded.

[2] Industrial Centres such as Syros, Edessa, etc.

were able to adapt to the post-war economic restructuring process, and to continue their operations until the beginning of the twenty-first century. However, the changes that marked the globalized economy in the late twentieth and early twenty-first century contributed to the final decline of the textile sector and the closure of almost all factories in the area.

The Ottoman period

The first spinning mill plant in Naoussa, Longos-Kirtsis-Tourpalis, was founded in 1874 (Fig. 1). Then followed, in the early 1890s, the establishment of Bilis-Tsitsis & Co., the construction of the Goutas-Karatzias factory in 1903, and the factory of Gr. Tsitsis & Co, in the first decade of the twentieth century, when the wool mills of Hadjilazarou & Co (Eria) and Lanaras-Pehlivanos were also founded[3] (Tab. 1).

Fig. 1. Longos-Kirtsis-Tourpalis at the beginning of the 20th century. Source: Municipality of the heroic City of Naoussa.

[3] Oikonomou 1990: 307–351.

Tab. 1. Naoussa's textile companies about 1911. Source: Oikonomou A., Naoussa's industry in 19th and 20th Century. In Naoussa 19th–20th century, (pp. 307–352). Cultural Society of Naoussa Anastasios Michail Logios. The dates in brackets refer to the factory's actual operation date.

Company	Establishment date
Longos-Kirtsis-Tourpalis	1875
Bilis-Tsitsis & Co	1891
Goutas-Karatzias & Co	1890 (1903)
Gr. Tsitsis & Co	1895 (1911)
Wool mills Eria SA (Hadjilazarou & Co)	1907
Wool mills Lanara-Pehlivanos	1910

The causes of Naoussa's textile industrialization have been researched in various treatises[4]. In 1822 the town suffered a great disaster because of its participation in the Greek War of Independence. The resettlement of the town was supported by the granting of economic privileges, while the general process of the Ottoman Empire's modernization, with laws which granted equal rights to Christians and Muslims[5], helped the formation of a local Christian trade and landholding caste. Its members, the so-called «chorbaji», were able to control the water supply from the waterfalls, necessary as the energy source of the factories. Initially utilizing the progress of water wheel technology, and later the water turbine, they were able to use the so-called «white coal» energy, exploiting at the same time the cheap labor of Macedonia's region[6]. The entrepreneurs turned to the cotton spinning industry, aiming to sell yarn to the enlarged market of domestic handicraft in the central Balkan region[7], also taking advantage of their easy access to cotton production areas[8]. At the same time, the contracts for the Ottoman military clothing manufacture

[4] Stougiannakis 1911: 145–157, Lapavitsas: 2010, Oikonomou op. cit., et al.

[5] Such as Tanzimat (1839) and Hati Humayun (1856).

[6] Hadziiosif 1993: 91, the male wage was 2 francs, while the female wage was 0.60 francs.

[7] Ibid, refers to potential markets the provinces of Thessaloniki and Bitola, while Palaskas 1986: 30–31, refers to the markets of Albania, Bosnia, Epirus, and Croatia.

[8] As the flatlands of Serres, Veria, Almopia et al.

helped the development of the local wool cloth industry[9]. This industrial activity was assisted by the new railway line linking Thessaloniki and Bitola in 1894, which facilitated trade.

The capital accumulation needed for the foundation of the local textile industry was developed by a financial interest alliance, often founded on intermarriages between the farm owners and the town's local or expatriate business families[10] (Fig. 2). It is no coincidence that all the industrial enterprises which were founded in Naoussa before World War II had more than one shareholder, with representatives of at least two families in their share capital. The investments of the local capitalists in Edessa and Thessaloniki resulted in their taking control of the textile sector of the European part of the Ottoman Empire[11].

Fig. 2. Family photo of Tsitsis, Platsoukas, Sefertzis, Longos, Tourpalis families about 1920. Source: Municipality of the heroic City of Naoussa.

The rapid industrialization of Naoussa, led to the creation of a local industrial tradition, a specialized workforce, and the constant

9 Lapavitsas op. cit: 40.

10 Based at the Greek state, Alexandria, Bosnia, etc.

11 Roupa and Hekimoglou 2004: 8–11, 7 out of 19 Christian business families were from Naoussa, while Gr. Tsitsis spinning mills in Edessa was founded from entrepreneurs from Naoussa.

recruitment of a low-paid female labor force from the surrounding rural areas. These changes resulted in social rivalries in the beginning of the twentieth century, and shaped the town's economy and society for the following years[12].

The period 1913–1940

The incorporation of Naoussa into the Greek state after the Balkan Wars in 1912 was followed by a ten-year-long war[13]. The local companies had to adapt to a different institutional framework, to face the shrinkage of their internal market, and to find new sources of raw materials. The industry of Naoussa was still important for the regional and national economy, due to the reduction of imports caused by the economic crisis of 1929. However, several companies had outdated machinery, while the economic crisis resulted in civil unrest and contributed to the over-indebtedness of many notable business families[14]. Many factories operated seasonally, while others completely suspended their operations[15] or changed their ownership structure[16]. Bilis-Tsitsis SA, the successor company of Bilis-Tsitsis & Co, relocated its headquarters to Thessaloniki in 1925. Despite these difficulties, new industries were established, such as Lanaras-Goutas in 1921, the first plant off the Arapitsa river, Lanaras-Kirtsis Wool cloth industry in 1922 (Fig. 3), and the Pennie Wool spinning mills-Lanaras Bros SA in 1928 (Fig. 4).

---

[12] The «chorbaji» and the so-called «popoulo» (the official name was the People's Association of Saint Demetrius), which was an alliance of part of the bourgeoisie with the poor landless farmers and the industrial workers, argued about the informal use of the church cash reserves as a source of credit for the local entrepreneurs, due to the lack of an organized banking system, and about other social issues. Apostolou 1986, Lapavitsas 2003, Valsamidis 1980.

[13] The two Balkan wars 1912–1913, Word War I 1917–1918, and Asia Minor war 1919–1922.

[14] Oikonomou op cit: 348; they used to grant each other personal guarantees for their business loans. This fact interconnected the fate of their businesses with their own family property, which resulted in a gradual loss of the companies' management.

[15] As in the case of the factory of Longos-Kirtsis, Tourpalis after a devastating fire in 1936, and Eria.

[16] As in the case of the Goutas-Karatzias factory in 1930, which was renamed Estia, while Lanaras-Pehlivanos was merged with Lanaras-Goutas.

Fig. 3. Lanaras-Kirtsis factory in 1922. Source: Municipality of the heroic City of Naoussa.

These ruptures in the economic life of the town were accompanied by continuities (Tab. 2), such as the Greek government's commissioning of military clothing contracts to Eria, and to the new wool cloth company of Lanaras-Kirtsis, though this was not an easily accomplished task. The two industries threatened the government with shutdown in 1924[17], and finally in 1926 they succeeded in sharing the state contracts with the Greek capital's companies[18]. These events helped the emergence of Lanaras family in the business world of Naoussa.

17 Tousimis 1996: 113–117, Hekimoglou 2015: 59–62.

18 Hadziiosif op. cit: 114.

Tab. 2. Naoussa's textile factories about 1939. Source: Oikonomou A.G., ibid.

Company	Condition of the factories
Longos-Kirtsis-Tourpalis	Non operational-Burned down
Bilis-Tsitsi SA	Operational
Estia (former Goutas-Karatzias)	Partially operational- Rented machinery to Pennie wool spinning mills
Gr. Tsitsis & Co	Partial operational
Eria SA	Non operational
Lanaras-Pehlivanos	Partially operational
Goutas-Lanaras	Partially operational
Lanaras-Kirtsis	Operational
Pennie Wool spinning mills-Lanaras Bros	Operational

The four sons of Christodoulos Lanaras established two large factories with modern machinery in Naoussa in the 1920s, and cooperated with members of Naoussa's bourgeoisie who had already settled in Thessaloniki and Athens, appearing as the leaders of the textile industry of Macedonia. Following the example of many provincial industries during the interwar period, they relocated their headquarters and founded factories in the Greek capital in order to be closer to the political decision-making area, and to meet more easily the needs of the largest consumer centre of the country[19].

[19] Leonidakis 2017: 20–34.

Fig. 4. Pennie Wool spinning mills – Lanaras Bros SA about 1930. Source: Municipality of heroic City of Naoussa.

The occupation period and the first post-war years 1941–1962

The Nazi occupation of Naoussa during World War II lasted from April 1941 to September 1944. During the war, several factories continued their operations, some of them concluding large contracts for the production of the occupation army's military clothing[20]. After the war was over, the Greek Civil War broke out, between 1946 and 1949, resulting in a turbulent political climate which included labour strikes, political assassinations[21], and repeated attacks on the town by the Communist Democratic Army[22]. The attack of 11-01-1949 was successful, and during

---

20 Manousakis 2014: 300, there are references to invoices concerning the Lanaras-Kirtsis factory (377.333, 33 RM) and Kokkinos (Gr. Tsitsis & Co) factory (313.416.67 RM).

21 Apostolou 2008: 36–37, 43, Karanasios 2004: 161–162, 183–186.

22 Ibid. pp. 194, 204, 243, 285. The Democratic Army attacked Naoussa on 30/06/1946, 01-02/10-1946, 18-19/08/1947, and 22-23/12/1948.

the three-day occupation, all the factory facilities were burned and blown up[23](Fig. 5).

Fig. 5. Pennie Wool spinning mills-Lanaras Bros SA after the destruction, January 1949. Source: Απ 'εδώ πέρασαν οι συμμορίται, The bandits passed through here (1949). Information Office, C Greek Army Corps.

The possible consequences of this industrial destruction, such as a rise in unemployment and in social tension, in the midst of the Greek Civil War in a region bordering the «Iron Curtain», led to the introduction of a new «bandit-stricken» industries' section among those funded by the Marshall Plan[24] (Tab. 3). Half the section's approved loan applications came from Naoussa's industries[25], the damage to which far outstripped

---

23 Margaritis 2002: 187.

24 The communist rebels were called disparagingly «bandits». Trikala, Karditsa, Karpenissi and Kalavryta were also labeled as "bandit-stricken towns".

25 There were 14 approved applications.

those of other towns; their approved loan amount accounted for 82 % of the total amount. Finally, seven companies from Naoussa[26] received $ 3,600,000 or 92 % of the total disbursements[27], on very favorable loan terms[28].

Tab. 3. Loan amounts to «bandit stricken» textile industries in Naoussa, 30-08-1954. Source: PIOP Folder CLC, «American Aid's Loans CLC and AMAG according to the accounting reconciliation of 30-08-1954» FOA2/SE2/SS9/FI2.

Company	Amount in dollars
Eria SA	877,100.9
Gr.Tsitsis &Co	716,261.1
Lanaras-Kirtsis SA	709,652.2
Pennie Wool spinning mills-Lanaras Bros SA	458,263.9
Bilis-Tsitsis SA	445,689.8
Vetlans & Co (former Goutas-Lanaras)	144,652.2
Sum	3,351,619.9

Under these terms and conditions, the reconstruction loans could be used only for reparations and reconstruction of the factories' building facilities and the renewal of the machinery, not for supplementary working capital for the purchase of raw materials. This resulted in the renovation of the facilities and the modernization of the factories' machinery, but also led to a delay in the companies' recommencement of their commercial production-sales cycle, as many entrepreneurs could not, or would not, contribute to their company's working capital. Moreover, the reconstruction loans were concluded with a dollar clause, thus the grant was in drachmas, but was repaid at the official dollar/drachma exchange

26 6 textiles and 1 rope manufacture.

27 This amounted to $ 3,900,000. PIOP Folder CLC «Loans of American Aid, CLC & AMAG based in the accounting reconciliation of 30-08-1954» FOA2/SE2/SS9/FI2.

28 The loans had a 15 year repayment schedule, the first 3 years had only interest payments with 1 % interest, and the following 12 years had 5 % interest.

rate[29]. When the drachma was greatly devalued in April 1953, the debt was doubled.

The textile industry in Greece during the 1950s experienced a major restructuring crisis (Tab. 4), due to the companies' over-indebtedness, the partial deregulation of imports, and an overproduction crisis. In Naoussa specifically, there was rising unemployment and continuing uncertainty about the local industry's future[30]. Many articles in the local press of the period dealt with the major economic and social problems created by the early deindustrialization of the town, demanding state support. A major blow for Naoussa, a town of 12,874 people at the time, was the gradual dismissal of the 1700 workers of the Lanaras-Kirtsis factory[31]. Despite the state support and its exporting activity, the company was dissolved, due to its increased debt obligations[32]. Its facilities and machinery were leased and the company continued its operations under the name of Pella-Olympos.

Regarding the rest of the companies, some like Eria[33] and Gr. Tsitsis & Co were dissolved, others such as Bilis-Tsitsis SA continued their operations on behalf of third textile industries[34], and Estia was sold and renamed Yfen. Nevertheless, Vetlans escaped insolvency by successfully changing its production to acrylic blankets. Pennie Wool spinning mills also changed its production to cotton yarn, while at the same time, a partnership of some employees of Gr.Tsitsis leased its facilities and equipment, and continued the yarn production[35].

---

29 The devaluation by 50 % meant that the loans should be repaid at a rate of 30,000 drachmas/dollar instead of 15,000 drachmas/dollar.

30 This was despite the gradual development of the agricultural fruit packing industry in the Greater Vermio area.

31 PIOP Folder Lanaras-Kirtsis «Audit Reports» FOA2/SE2/SS4/FI43001/SFL8.

32 It is indicative that in the last debt settlement to which it was subjected in 1960, the company had debts of about $ 6,533,000, a current value of about $ 54,000,000.

33 This was despite the granting of the largest reconstruction loan ($ 877,100), and an extra 200,000 $ working capital.

34 PIOP Folder Bilis-Tsitsi SA, «Audit Reports», FOA2/SE2/SS4/FI43005/SFL13.

35 Samara 1994: 30.

Tab. 4. Naoussa's textile factories about 1962. Source: PIOP Folder Bilis-Tsitsis SA «Audit Reports», FOA2/SE2/SS4/FI43005/SFL13.

Company	Condition of the factories
Longos-Kirtsis-Tourpalis	Abandoned
Bilis-Tsitsis SA	Operational
Yfen (former Goutas-Karatzias)	Operational
Gr. Tsitsis & Co	Partially operational, into liquidation
Eria SA	Rented by Vetlans, into liquidation
Vetlans & Co (former Lanaras-Goutas)	Operational
Lanaras-Kirtsis	Non operational, into liquidation
Pennie Wool spinning mills-Lanaras Bros SA	Operational

From development to decline 1963–2012

The Naoussa's spinning mills SA were founded in 1963 as a joint venture between a public development organization[36] and Pennie Wool spinning mills with industrial facilities in Naoussa but its headquarters based in Athens. The main business goal of the new company was the export of cotton yarn[37]. In addition to the existing facilities, a new large spinning plant was built just outside the town. Despite the initial difficulties the company's growth had been striking, as it arranged a vertical integration of its production, by setting up a number of affiliates[38], while at the same time, it expanded its industrial facilities, operating eight factories in the late 1990s in the greater Naoussa region[39]. Vetlans also expanded its industrial activity, because of its major export performance until the mid-1980s. It established a new industrial plant[40] and acquired the

---

36 The Organization for Industrial Development.

37 PIOP Folder, Naoussa's Spinning Mills «Report for the application for plant assets » FOA3/SE6/SS3/FI3P111/SFL5. The Organization for Industrial Development contributed with 60 %, while Pennie Wool Mills with 40 %.

38 Such as a knitting company (Trikolan in 1973), a special yards spinning mill (Olympic Textiles in 1978), and readymade garments companies (Verlan in 1978 and Fanco in 1985).

39 In Stenimacho, 7 km outside Naoussa, where a small industrial area was built, in Episkopi, 12 km, and in Loutrochori, 15 km from Naoussa.

40 In Agios Athanasios, Thessaloniki.

former Eria and Pella-Olympos factories[41], at the same time relocating its headquarters to Thessaloniki. In 1964 a group of redundant workers of the ex-Gr. Tsitsis Co plant, established the Varvaresos & Co partnership, while in 1974 Varvaresos SA was founded. The company built a second plant outside of Naoussa in order to meet the export demand[42].

It is evident that the main energy source of the modern factories was not hydroelectric, and the headquarters of the largest companies were based either in Athens or in Thessaloniki. On the other hand, many smaller textile companies were operating at that time in the area of Naoussa[43]. This local development, which proved the increasing demand for Greek textiles, was part of the broader expansion of the Greek economy which had been steadily growing from the early 1960s to the mid-1970s[44]. Many companies took advantage of the incentive laws for the provincial industry, and those favoring exports, exploiting at the same time the lower Greek labour cost[45], and the international economic stability. The situation, however, would change with the turbulent economic conditions of the 1970s. The local companies could not easily compete with the lower consumer prices of the Southeast Asian textiles which gradually, from the 1980s, entered the Western European markets[46]. The opening of the borders of the Balkan countries[47], the growing textile production of Turkey[48], China's accession to the WTO in 2002, and the abolition of the multi-fibre arrangement in 2005[49] would complete the process of emerging competition from countries with lower labour costs. Moreover, domestic factors such as increasing labour cost,

---

41 'IX Shareholder General Meeting of Vetlans SA, 30-06-1978, Archive of Naoussa's Industrial Heritage Documentation Center, the company was already the major shareholder of the former Lanaras-Kirtsis plant.

42 Samara, op. cit: 30–33. The second plant was built in Stenimacho's industrial area.

43 There were at least 10 smaller textile companies operating in Naoussa at that time.

44 World Bank. https://data.worldbank.org/indicator/NY.GDP.MKTP.KD.ZG?locations=GR, last view 20-07-2018. Greece had an average nearly 8 % annual GDP growth between 1961 and 1973.

45 The workers' demands had been suppressed due to the intense political climate of post-war Greece, resulting in lower wage rises.

46 This came about with European Regulations 1440/1980 and 743/1985.

47 Much of Greek manufacturing production would be transferred there in the 1990s.

48 The Customs Union between Turkey and the EU came into effect on 31/12/1995.

49 The MFA (1974–2004) imposed quotas on the amount of textiles that developing countries could export to developed countries.

the over-financialization of the economy, and inflexible manufacturing production, intensified the problems (Fig. 6).

The first textile company to shut down was Yfen in 1984. Vetlans followed, closing first its own factory and the former Eria in the early 1990s, and finally Pella-Olympos, in the mid-1990s[50]. Many smaller companies also ceased their operations. The Klonatex Group, as the Lanara family's Naoussa's spinning mills were renamed, exploiting the rise of the share prices of the Greek Stock Exchange at that time, bought off at least 25 textile factories[51], and expanded its activities in the non-textile sector. Despite state and bank support, the Stock Exchange collapse after 1999 resulted in the gradual closure of the company's factories in Naoussa, starting from 2004, and finally its bankruptcy in 2012[52]. Today there are no industrial facilities inside the town of Naoussa. The only textile company that survived, Varvaresos SA, operates two plants in the industrial area just outside the town, with 226 employees[53].

Fig. 6. One of the abandoned factories of Naoussa's spinning mills. Source: Author's photograph.

50 The situation worsened due to the wooden box manufacturing crisis.

51 This occurred not just in Greece but also in the greater Balkan region.

52 The estimate of Klonatex workers in greater Naoussa was about 800.

53 https://varvaressos.eu/gr/, last view 20-07-2018.

Towards the future

The deindustrialization caused massive unemployment in Naoussa, which resulted in growing civil unrest such as protests and strikes, and in internal and external migration. The government response to this economic restructuring crisis was the granting of an early retirement program to a portion of the redundant workers, as a benefit, in order to provide them with financial support[54]. In the meantime, the emergence of the Greek economy crisis in 2009 reduced the state funds which could facilitate the transition of Naoussa's region to a diversified economy.

Throughout the early 1990s and beyond, new perspectives of the city have been sought. The proposed alternatives were the development of agriculture and tourism, as the industries of fruit packing and box manufacturing also experienced financial problems. Although the flatland of Naoussa and the surrounding hills are especially fertile, producing the well-known Naoussa wine, and a large quantity of peaches, cherries and other tree crops, it is difficult to build a sustainable economy based only on agriculture for a town of about 20,000.

The greater Naoussa region is endowed with natural beauties which are mainly capitalized for winter tourism[55], while in spring the Carnival of Naoussa is more widely known. An effort is being made by the municipality[56] and private entities to extend the tourist season through the utilization of the industrial heritage, while at the same time highlighting the historical past of the city, and giving prominence to wine tourism[57].

The capitalization of the buildings' wealth left behind by the deindustrialization started as early as 1989. The oldest industrial building of Naoussa, the Longos-Kirtsis-Tourpalis factory, after 15 years of restoration, initially hosted the university department of Naoussa. The renovated Eria building hosts the Municipal Centre for Industrial Heritage Documentation, while the Vetlans building houses municipal offices

---

54 L3460/2016, article 13. It did not provide future employment alternatives to those who were not covered by these provisions.

55 The ski resorts of 3–5 Pigadia and Seli are situated 17 km and 20 km away from Naoussa respectively.

56 The municipality of Naoussa aims to display information signs accessible by mobile phones & tablets through QR technology.

57 The local traditional winery has established the PDO xinomavro, for the wine of the region.

(Fig. 7). The building of Bilis-Tsitsis is abandoned, as well as the Lanaras-Goutas old factory. The old Yfen plot is used for hydroelectric production; the same use is intended for the Gr. Tsitsis Co plot, while both buildings are abandoned. The Pella-Olympos plant (former Lanaras-Kirtsis) has no use, although it is in good condition (Fig. 8), while Trikolan (formerly Pennie Wool mills) is used as a warehouse for stock sale[58] (Tab. 5).

Tab. 5. Naoussa's textile factories in 2018. Sources: Newspapers, Municipality of heroic City of Naoussa.

Company	Condition of the factories
Longos-Kirtsis-Tourpalis	Renovated
Bilis-Tsitsis SA	Non-operational-abandoned
Yfen (former Goutas-Karatzias)	Non-operational since 1984 – abandoned
Varvaressos SA (former Gr. Tsitsis & Co)	Non-operational since 2007-abandoned
Vetlans SA (former Eria)	Renovated- Center of Industrial Documentation
Vetlans SA (former Lanaras-Goutas)	Non-operational since 1990-abandoned
Vetlans SA	Renovated- Municipal Library-Municipal services
Pella-Olympos (former Lanaras-Kirtsis)	Non-operational since 1998
Trikolan (former Pennie Wool spinning mills)	Non-operational since 2004-Warehouse

One of the Naoussa's spinning mill plants is sufficiently equipped with modern machinery, and could operate again, if there was investment interest[59]. The fact is, however, that as time goes by, such projects are more difficult to implement, and they are treated with a lot of scepticism especially by younger people, who are less affiliated to the textile industry.

58 Information from the Municipality of Heroic city of Naoussa.

59 It has vertically-integrated yarn production, knitted cloth, and garment cutter plant, well-equipped with modern machinery.

Fig. 7. Vetlans building. Source: Author's photograph.

Fig. 8. Pella-Olympos- former Lanaras-Kyrtsis building. Source: Author's photograph.

Conclusions

This short chapter attempts to picture some aspects of the textile industry in Naoussa, and its changes throughout the twentieth century. The local industry was important not only for the town itself but also for the regional and national Greek economy. The business class which contributed to the development of Naoussa gradually moved to Athens or Thessaloniki, taking with it the headquarters of their companies, but maintaining some of the local industrial facilities. The final downturn of the local textiles followed the general trend of the decline of traditional industries all over Europe, caused by the industrial maturity of the developing countries, and the trade globalization. Although some local factories were adequately modernized, the potentials of producing high quality products with reduced consumer prices could not be realized, due to the special features of the Greek entrepreneurship at the end of the twentieth century, and an inadequate government response to a structural textile crisis. This led to the Naoussa's community struggle for a diversified economy in the new post-industrial era. The development of qualitative agricultural production, combined with the growth of all-year-round tourism could strengthen the resilience of the town. This economic model could function better if there was a coherent plan for the inclusion of the former factories (Fig. 9) in the city's tourist map, and the improvement of the infrastructure of the city's river footpath, as the majority of the industrial sites are situated alongside the river Arapitsa. The lack of funds and the complicated ownership of some industrial buildings did not yet allow the completion of this promising investment either by the municipality of Naoussa, or by private investors. The usage of some former factories as an alternative water power energy source, or as municipal services buildings, functions supplementary with their diversified post industrial use. Regarding the commodification of the town's heritage, the defunct textile factories could be used, not for just another one industrial theme park, or as a reminder of the town's former economic growth, but as an example of the inclusion of the recent past in the town's collective memory, which could help the citizens of Naoussa to come to terms with the town's recent deindustrialization. In conclusion Naoussa maintains many future competitive advantages, but the process of its economic paradigm shift is underway, and the final results are yet to be seen.

Fig. 9. Textile factories situated inside the town of Naoussa. Source: The map comes from the website http://zafirakishc.blogspot.gr/2011/07/blog-post_24.html. The author has cited the factories' location.

Bibliography

Archival material

Historical Archive of Piraeus Bank Cultural Foundation (PIOP)

Archive of Naoussa's Industrial Heritage Documentation Center

Rizospastis (newspaper)

Kathimerini (newspaper)

Macedonia (newspaper)

Voice of Naoussa (newspaper)

Embros (newspaper)

References

Apostolou, S. (1986). Ο λαϊκός σύνδεσμος Ναούσης «Άγιος Δημήτριος» και το ιδεολογικό του υπόβαθρο [The People's Association of Naoussa and its ideological background]. *Niaousta* 34, 5–12.

Apostolou, S. (2008). *Η Νάουσα στην δίνη του Εμφυλίου Πολέμου*, [*Naoussa in the Whirlpool of Civil War*]. Self-published.

Hadziiosif, C. (1993). *Η γηραιά Σελήνη, η Βιομηχανία στην Ελλάδα 1830–1940*, [*The Old Moon, The Industry in Greece 1830–1940*]. Themelio.

Hekimoglou, E. (2015). *ΣΒΕΕ 1915–2015 τα πρώτα εκατό χρόνια* [*SBBE, 1915–2015, the first 100 years*]. SBEE [Association of the Industrialists of Northern Greece].

Karanasios, C. (2004). *Η Νάουσα στην Εθνική Αντίσταση και στον Εμφύλιο* [*Naoussa during the National Resistance and the Civil War*]. Self-published.

Lapavitsas, K. (2003). Η εκβιομηχάνιση της Νάουσας κατά την ύστερη οθωμανική περίοδο [The industrialization of Naoussa in the late Ottoman period]. *Niaousta*, 102–103, 22–27.

Lapavitsas, K. (2010). Βιομηχανική Ανάπτυξη και κοινωνικός μετασχηματισμός στην Οθωμανική Αυτοκρατορία [Industrial Development and Social Transformation in the Ottoman Empire]. In K. Lapavitsas (Ed.), *Ισχνός Καπιταλισμός, η Μακεδονία κατά την ύστερη Οθωμανική περίοδο* [*Fragile Capitalism: Macedonia during the Late Ottoman Period*], (pp. 21–76). Novoli.

Leonidakis, N. (2017). *Η χρηματοδότηση μιας «συμμοριόπληκτης» βιομηχανίας μέσω του Σχεδίου Μάρσαλ, η περίπτωση των Εριοκλωστηρίων Πεννιέ- Αφοι Χρ.Λαναρά ΑΕ* [*The financing of a «bandit-stricken» industry, the case of Pennie Wool spinning mills»*], [unpublished master's thesis]. University of Crete.

Manousakis, V. (2014). *Economy and Politics in World War II Greece*, [unpublished doctoral dissertation]. Aristotle University of Thessaloniki.

Margaritis, G. (2002). *Η Ιστορία του ελληνικού εμφυλίου πολέμου1946–1949*, [*The history of Greek Civil War 1946–1949*]. Vivliorama.

Oikonomou, A. G. (1990). Η Βιομηχανία στη Νάουσα από το 19° στον 20° αιώνα [Naoussa's industry in 19th and 20th Century]. In *Νάουσα 19ος-20ος αιώνας [Naoussa 19th-20th century]* (pp. 307–352). Cultural Society of Naoussa Anastasios Michail Logios.

Palaskas, G. (1986). Οι πρώτες βιομηχανικές εγκαταστάσεις στην Κεντρική Μακεδονία 19ος-20ος αιώνας [The first industrial facilities in central Macedonia in 19th – 20th century]. *Archaeology* 18, 29–36.

Roupa, E., & Hekimoglou, E. (2004). *Η Ιστορία της επιχειρηματικότητας στη Θεσσαλονίκη, η επιχειρηματικότητα στην περίοδο 1900–1940 [The history of entrepreneurship in Thessaloniki 1900–1940]*. Cultural Foundation of Northern Greece's Entrepreneurs.

Samara, L. (1994). Βαρβαρέσος Α.Ε., Νηματουργεία Ναούσης [Varvaresos SA, Naoussa's spinning Mills]. *Niaousta* 66, 30–33.

Stougiannakis, E. I. (1924). *Ιστορία της πόλεως Ναούσης [History of Naoussa town]*. G&E Stougiannakis.

Stougiannakis, E. I. (1911). Νάουσα [Naoussa]. *Macedonian Calendar* (D), 145–157.

Tousimis, G. (1996). Σελίδες από την εργατική ιστορία της Νάουσας [Pages from Naoussa's labor history]. *Niaousta* 76–77, 113–117.

Valsamidis, E. (1980). Ο κανονισμός του Πούπουλου [The charter of popoulo]. *Niaousta* 10, 6–8.

Vasradvellis, I. K. (1967). *Οι Μακεδόνες κατά την επανάστασιν του 1821*, [The *Macedonians during the Revolution of 1821*], Society for Macedonian Studies.

Interviews

Valsamidis, A., Lawyer. Interview by author. Naoussa, 05-06-2018.

Valsamidis, S., Deputy Mayor of Culture of Heroic city of Naoussa. Interview by author. Naoussa, 05-06-2018.

Ruin and time, abandoned settlements. The small town of Granadilla

F.-Javier Ostos-Prieto and
José-Manuel Aladro-Prieto

Introduction

Founded in the ninth century and feudal in origin, Granadilla is a small settlement located in the north of the province of Cáceres, in the Spanish region of Extremadura. Currently, it is an abandoned and deserted settlement without inhabitants. In the 1960s, the state carried out an expropriation, due to the Gabriel y Galán reservoir construction in the Alagón River. As everything near the river was under water, including the Roman bridge, the same situation was expected with the town as happened with other reservoirs in Spain. However, this settlement was saved from the waters thanks to its geographic height. In spite of that, and as a consequence of this situation, its inhabitants were forced to migrate to other nearby villages. Granadilla was completely deprived of its territory, isolated in a small peninsula above the new reservoir. Parallel with the appearance of the reservoir, all the water edge was replanted with new tree species, pines and eucalyptus, which left Granadilla hidden and forgotten between the water and the vegetation.

Despite its condition, in 1980 Granadilla was declared an Historic-Artistic Site by the Spanish state. Owing to this, the rehabilitation and recovery of the most characteristic elements began, as well as some houses. Only twenty percent was restored, essentially the important elements: the castle, the walls, the church, the houses in the main square and streets; leaving the remaining part dilapidated. Four years later, Granadilla together with the towns Búbal (Huesca) and Umbralejo (Guadalajara) were selected for the *Programa de Recuperación y Utilización Educativa de Pueblos Abandonados* [Recuperation and Educational Utilization

Programme of Abandoned Settlements]. The objective of this program is the recovery and consolidation of Granadilla by communal living and study by secondary school students. These students come from all over Spain to stay there for seven days. That is the only activity developed in the place, which is here analysed under a contemporary heritage view. Also, it is important not to forget the evaluation of its romantic perspective, which comes from its ruined state and the exceptional artificial site.

This chapter presents and analyses the unique case of the heritage in Granadilla. There is though as well the possibility or capacity of this heritage to address the recovery and rehabilitation of a complete urban structure, without recovering its urban condition. The current situation has been studied through the analysis and extraction of documentary, literary and graphic sources, and the knowledge and understanding of the process suffered in Granadilla. Additionally, the fieldwork carried out enables precise recognition of the state of rehabilitation and ruin. Finally, knowledge of the entire process allowed an evaluation of the current romantic attraction of the ruin in the twenty-first century: the power of what was destroyed on what was built, as well as the survival of the reconstruction of the scenery that takes place in Granadilla.

A remote place

Within the European panorama and landscape of the west of the continent, Spain stands out for the dimension of its heritage; within its limits, to the west and occupying a central position, is the province of Cáceres. According to the territorial organization of the country, it belongs to the Region of Extremadura, outstanding in the Spanish panorama for its rich architectural, natural and intangible heritage.

The small town of Granadilla is located in the north of Cáceres, in the surroundings of the Natural Park of Las Batuecas-Sierra de Francia. It is in an exceptional location on the edge of the Gabriel y Galán reservoir. Its proximity to the Roman roadway called *Via de la Plata*[1] has

1 The Iberian Peninsula had two important roads that articulated the territory in Roman times: the *Vía Augusta*, which linked Cádiz with Rome (East-West), and the *Vía de la Plata*, which connected Seville with Astorga (North-South), in Muñoz Hidalgo 2010: 8–9.

historically given it an important territorial position. Although uninhabited, its peninsular location in the marsh on a hill enhances it above the landscape of the Valley of the Alagón River (Fig. 1).

Such a striking landscape position is the result of a deep anthropogenic alteration of the landscape with the construction of the reservoir. When it appeared, the condition of the settlement changed forever, and its new and picturesque geographical situation paradoxically led to its total depopulation.

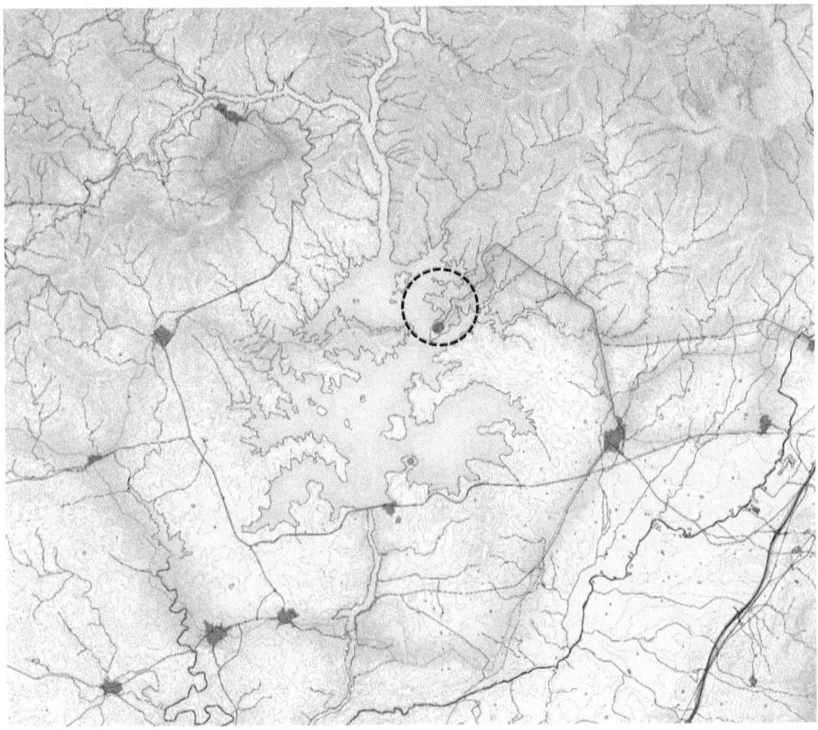

Fig. 1. Granadilla location in Gabriel y Galán reservoir. Source: Made by F. Javier Ostos Prieto based on CAD – SITEX Extremadura.

Granadilla, stronghold-settlement

Although there was a small settlement with Muslim origin dating from the ninth century, the appearance of the town of Granada[2] took place in the middle of the twelfth century. The foundation was in the context of the peninsular wars between the Christian and Muslim kingdoms. Its repopulation was ordered by King Fernando II, placing the town as a defensive point on the border of the Christian Kingdom of León. Its border location and the connectivity of the territory in the north-south and east-west directions were perfect for the fortified town.[3] Despite its Christian foundation, a decade later it was occupied again by Muslim troops. This new stage inherited the morphology of its urban layout. The remains are found in its fortifications and its walled structure. Once again in Christian hands, its appearance barely changed, except for the church introduced into the urban scene.

The village quickly reached large territorial importance. Politically, it obtained representation in the courts of the kingdom during the reign of Alfonso XI. It became the administrator of a multitude of smaller towns and was the center of political and territorial power in the whole northern region of Extremadura. It was head of the province and manager of different towns. Villages and hamlets as far away as La Alberca and Sotoserrano, in the current province of Salamanca, came under its influence, as well as others closer to it such as La Zarza de Granadilla, La Granja, La Pesga, Ahigal, Aldeanueva del Camino, Mohedas, Pinofranqueado or Nuñomoral. In the middle of the twentieth century, the appearance of the reservoir fractured the territorial structure. The municipal district was divided, its administration was transferred to other settlements, and the town of Granadilla was evacuated.

Today, the fortress-town remains enclosed within its walls, on the slate mound, surrounded by water, hidden between pines and eucalyptus trees. Despite its importance, the place never exceeded the fortified limits, being confined in a walled space. The growth forecast was much higher, which allowed a generation of vegetable gardens inside the small town.

---

2 The name of the town was kept until the conquest of Granada by the Catholic Kings in 1492. To avoid confusion between the two Granadas, the first one changed to "Granadilla". In Lozano Bartolozzi 1980: 54–65.

3 Caballero González 2004: 28–36.

The enclosure had four gates, where only two are still open in the wall. Due to its current location on the peninsula, only the one next to the castle is used. Among its most important constructions are those that are expressions of the military, ecclesiastical and civil powers: the castle, the Asunción church and the town hall, located in the main square. The castle stands out at the main entrance to the village giving a powerful feudal image. The church is located in the middle of the small town. Above it, its bell tower still calls to the mass every 1st November[4]. Finally, the town hall square constitutes the nerve and geographical centre of Granadilla (Fig. 2). With a semi-circular floor plan, this space, the main one in the town, holds municipal buildings such as the prison, the school and the water cistern.

The settlement dwellings were mainly one-storey houses with gabled roofs. Those located in the square and the main streets reached two floors. Although they lacked interior courtyards, most of them had poultry houses and storages. Structurally, they were made up of load-bearing walls and wooden beams, a common type of construction in the area. The slate of the site was the base of the masonry walls, which might or might not be covered with lime depending on the importance of the dwelling.

4 The inhabitants of Granadilla meet twice a year: on 1 November (All Saint's Day) and 15 August (Asunción Day, the saint of Granadilla) as a reminder and as a way of living together. Alvarado Corrales 2014: 25.

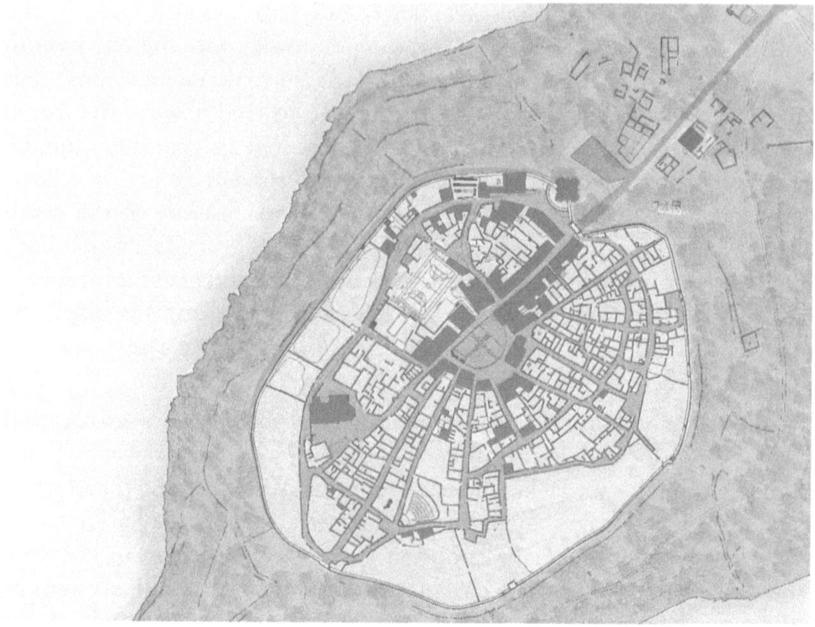

Fig. 2. Granadilla, current state. Restored buildings are marked with dark shading.

Hydraulic infrastructure and underwater heritage

The increase in population from the twentieth century onwards was widespread worldwide. In the case of Spain, the 18,830,649 inhabitants in 1900 had changed by the year 2000 to 40,499,791,[5] almost 2.5 times more in just one century. In the middle of the last century the gradual growth of the population, added to the usual scarcity of rainfall in much of the country, led to the proliferation of water infrastructures related to large-scale water storage. Rivers with a constant flow and located in rugged geographic environments were flawless for the placement of reservoirs.

In Spain, there are many reservoirs scattered throughout its peninsular territory. Andalusia and Extremadura stand out among the Spanish

---

[5] Retrieved from http://www.ine.es/welcome.shtml (31st July 2018).

Communities with the highest number and the highest quantity of water kept in them. Andalusia's storage capacity is 12,035 hm^3, while Extremadura's storage capacity is 14,219 hm^3. This places Extremadura in first place against the Region of La Rioja, the last with only 136 hm^3.[6]. The construction of this infrastructure flooded many settlements located in valleys. About 500 ghost towns are counted under the water. In years of drought, these towns return to view, emerging as if they were sunken ships. Their memory and origin have been forgotten in many cases, being remembered only by their older inhabitants.

The Gabriel y Galán reservoir appeared in this context in 1961, on the Alagón River. The name of the reservoir refers to the poet José María Gabriel y Galán, an inhabitant of the Alagón Valley[7]. The reservoir has a capacity of 925 hm^3 with a total dam height of 72 m. The construction of this large infrastructure aimed to provide abundant storage water for human and agricultural consumption, always as a key element of a plan to put more than forty-three thousand hectares under irrigation[8]. This large extension allows a common benefit of water to a large livestock sector.

Although the flooding of towns has happened throughout the Iberian Peninsula, some of the most striking examples have taken place in Extremadura. Talavera la Vieja (Cáceres) stands out in this region, where the reservoir of Valdecañas completely flooded the remains of the Roman city of Augustobriga in 1963. This specific case led the inhabitants of the city to move their most precious asset, the ruins of the ancient Roman temple. The columns and lintels were moved to the new shore of the reservoir in the current municipal district of Bohonal de Ibor[9]. In spite of the unavoidable flooding, the inhabitants sought to keep their heritage and identity safe.

Cases such as the example of Augustobriga instigate a debate on the question of flooded heritage. There is clear singularity of the element which is destined to be submerged, with clear examples at Spanish and international levels. On a world scale, the chief reference point is

---

6 Water storage capacity data obtained from the Spanish reservoir website. Retrieved from www.embalses.net (31st July 2018).

7 Fernández-Daza Álvarez 2005: 49–51.

8 Anonymous 1963: 47.

9 Santos 2014: Retrieved from https://www.traveler.es/naturaleza/articulos/lo-que-el-embalse-se-llevo-pueblos-espanoles-sumergidos-bajo-el-agua/6189 (31st July 2018).

undoubtedly the rescue operation of the Abu Simbel[10] temples in 1964. They faced the risk of being flooded by the waters of the Nile after the construction of the Aswan dam. The action was carried out by an international team and financed by UNESCO[11] [12]. This operation involved the movement of the temples 200 meters away on to land 65 meters higher. The objective was to avoid being covered by the new water level. They were moved piece by piece and then rebuilt at the new site. Thanks to this intervention, Abu Simbel was saved and now can be visited, while a multitude of heritage sites continue to be flooded in Spain.

Another similar intervention, very close to Granadilla, was the moving in 1982 of the Roman bridge of Guijo de Granadilla. It was built over the Alagón River and threatened by a smaller dam erected downstream. Due to the flooding, the loss of the original bridge, which was part of a historic road, forced its replacement by a concrete and steel bridge. The old original masonry infrastructure had to have a new location. Without its general function, an interesting recovery project located the old bridge over the reservoir, shown as an *objet trouvé* a few meters from its original location[13].

Every year of drought, forgotten pieces emerge from large water expanses. They could be famous Roman cities like Augustóbriga or significant architectural examples, such as the Romanesque church of the Campoo reservoir whose reappearance became popular in 2017. "When the reservoir dried up, the church was still there. Five decades after being flooded by the creation of the Aguilar de Campoo reservoir, in the north of Palencia, the town of Zalima and its church were brought to life."[14]. The emergence of a flooded heritage site usually enjoys high media coverage, being the subject of many articles, reviews in newspapers and news

---

10 The complex was composed of two temples: The larger one was dedicated to Ra, Ptah and Amon and the smaller to the goddess Hathor. In Siliotti 1994: 272–276.

11 In addition to the UNESCO funding for the relocation of Abu Simbel, a request was made for international assistance to safeguard other temples. The United States, Italy, the Netherlands, Spain and Germany provided their assistance in 1968. In compensation for their help, each country received a temple. Ibid. p. 270.

12 In the case of Spain, the Temple of Debod was granted, and moved to its new location in Madrid. In Almabro Basch 1971: 83–88.

13 Junquera and Pérez Pita 2003: 74–76.

14 Riaño 2017. Retrieved from https://www.elespanol.com/cultura/patrimonio/20170830/242976406_0.html (31st July 2018).

during the summer months. However, although there is general concern about the flooded heritage, increased by the media, there is no action; in fact, they are simply left to their destiny.

The state, which is bound by the constitution to ensure that its cultural heritage is safeguarded, is paradoxically the main promoter of the reservoirs, responsible for the destruction of part of the heritage. The construction of these infrastructures has been of vital importance to maintain the pace of progress of contemporary society, cooperating decisively in the improvement of the living conditions of the population and their territorial roots. In addition to the water supply, reservoirs play a crucial role in electricity production. The main driver of contemporary developments is at the same time a source of renewable energy. In the case of the Gabriel y Galán reservoir, hydroelectric development was inaugurated in the early 1980s.[15]. These are advantages which have sometimes led to the unavoidable uprooting of the inhabitants, supposed beneficiaries. As well as this, there is the loss of constituent elements of their cultural identity and our cultural heritage. These are local losses, sacrificed in favor of a solution to a global problem.

Moving the most relevant heritage elements has been the immediate solution to the problem generated, allowing the material condition of the cultural property to be safeguarded. Doubts arise when losses caused by the de-contextualization of the asset are noted. How much of its identity value really lies in its formalization and materiality? Could we move the most important buildings in Rome to different parts of the world to safeguard them from the pollution of the Eternal City without detriment to their universal value? Once the bridge of Guijo de Granadilla was lost forever, was it worth the economic effort invested in the preservation of its memory?

Granadilla, abandoned settlement

In June 1955 the Spanish state approved the expropriation, evacuation and abandonment of the entire town of Granadilla. It forced the immediate transfer of all its population to nearby towns, dispersing it throughout the territory.

---

15 Anonymous 1982: 7.

> "The families of Granadilla had to leave the houses where they lived and their parents and grandparents before them; they had to leave behind streets and experiences, see the water rise and flood the orchards, olive groves and pastures, even the cemetery, where they cried and remembered their people, whose remains they also had to move." [16]

The last family left the small town in 1965 when the hydraulic infrastructure work was completed and the flooding of the entire Alagón Valley began[17]. The river, the origin of the settlement in which Granadilla is located, was the same that caused its end and changed it forever.

Despite the abandonment caused by the imminent risk of flooding, Granadilla never disappeared (Fig. 3). It continued on the surface, on the hill that once defended it and now protects it from a new threat. Its inhabitants had to leave homes that were never lost. However, those places and buildings, such as the old bridge over the river or the cemetery, which were on lower levels, were hidden forever.

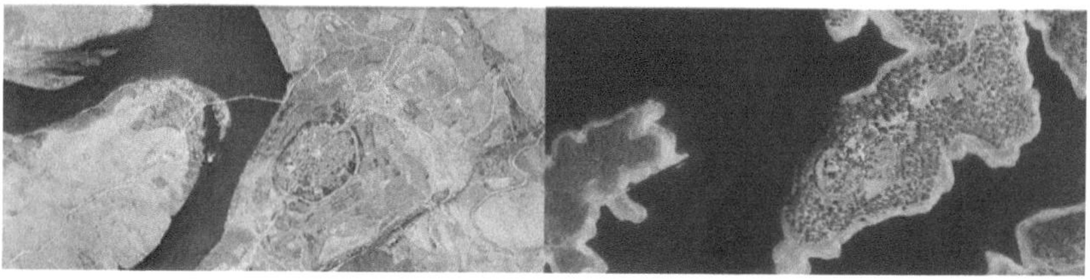

Fig. 3. Left. American Flight 1956 – Right. Air Flight 2017. Source: Spanish National Geographic Institute.

After the first observation that the walled enclosure was not flooded, the attractive landscape characteristics achieved by the town in its peninsular topography soon caught attention. Tourist proposals did not take long to appear. As early as 1963, the General Director of Tourism Promotion announced that a project for "the reconstruction and rehabilitation of Granadilla Castle [...] for a tourist hostel" had been accepted[18]. The idea was suggestive in a provincial interior lacking in tourist attractions. The lodging would generate an attractive new focus and the

---

16 Alvarado Corrales op. cit: 25.

17 Ibid, p. 24–25.

18 Gutiérrez Macías 1963: 13.

possibility of increasing the province's income. However, the proposal was not without a certain risk of cultural trivialization. In the end, it was not carried out, largely because its heritage potential began to be appreciated and as a result, a protection dossier was processed.

Although the small town could have been frozen after the abandonment, this did not happen. A pine and eucalyptus reforestation process took place along the entire bank of the new reservoir in the 1970s. The empty town served as accommodation for the various employees. During the winter, these workers burned a large number of wooden elements such as beams and wooden items to protect themselves from cold. This activity led the town to deteriorate, to the point of causing the total disappearance of many buildings. Faced with its imminent destruction, the alarms went off and the procedures for its protection as cultural heritage began. After a few years of study and data collection, a report was made in 1967 in which the heritage and historical value of Granadilla was stipulated in order to prevent its disappearance[19]. However, it was not declared an "Historic-Artistic Site"[20] until 1980. The declaration distinguished two different sectors: the "historic-artistic zone", bounded by the perimeter of the wall, containing the entire intramural wall, including the castle and the wall; and a "buffer zone" within a polygonal line outside the wall and at a constant distance of 800 m from it. This declaration entailed the immediate protection of Granadilla by the Spanish state.

From this year, and for the following three years, rehabilitation and repair works were carried out throughout the complex in order to paralyze its progressive deterioration and destruction (Fig. 4). The intervention was focused mainly on the key and unique elements of the population[21] [22] [23]. Pieces such as the castle, the walls and the church were the first to be restored, considered to have the greatest artistic value. Although apparently it was a restoration in style, the incorporation of contemporary materials and techniques (steel and waffle slabs) have caused a certain

---

19 Espejel Díez and Fernández de Castro 1967.

20 The legal declaration of the Historic-Artistic Site precedes the current law, which changed the term to Historical Site. In Ley 16/1985, de 25 de junio, del Patrimonio Histórico Español. Boletín Oficial del Estado, 155, de 29 de junio de 1985.

21 Espejel Díez 1995: 7–14.

22 Úbeda de Mingo 1985: 5–6.

23 Maldonado Ramos; Vela Cossío and Floriano Martín 1995: 5–50.

sense of confusion. The other strong point of the rehabilitation was the recovery of part of the old houses. Mainly, the perimeter dwellings of the central square were restored, including public buildings such as the town hall, as well as those houses located in the main street and around the castle (Fig. 5). The objective was not to recover all the buildings, but enough of it to enable it to be understood and to be able to accommodate future uses within its heritage limits.

Fig. 4. View of ruined houses with the church at the back.
Source: Photography taken by F. Javier Ostos Prieto (2017).

Those buildings, or their remains, that were not rehabilitated, were partly consolidated in their ruinous situation to avoid their definitive degradation. Even so, this process reached only a small part of the whole, keeping the loss of the last walls in much of the urban settlement.

Fig. 5. View of the Main Street and houses from the castle.
Source: Photography taken by F. Javier Ostos Prieto (2017).

Fortunately, in 1984, Granadilla was integrated into the Program for the Recovery and Educational Use of Abandoned Settlements[24] to host educational and training activities for schoolchildren in compulsory secondary education. This program is currently being carried out in three towns: Umbralejo (Guadalajara), Búbal (Huesca) and Granadilla (Cáceres). Its development has an annual character[25]. Every week students from different schools come to the village from different places and for seven days they live together in the village. The program aims to bring the new generations closer to the importance of rural life and to initiate them in their knowledge, in environmental education and in areas that are not learned in schools. The schoolchildren participate in workshops in agriculture, livestock farming, carpentry, ceramics, masonry, recycling or photography, in which they will put into practice new habits of healthy living. In addition, many of these tasks and jobs are focused on the conservation and preservation of the town.

Concerning the other towns included in the program, Búbal is in the north of the province of Huesca. This settlement was expropriated for the construction of a reservoir, which would eventually flood only the lowest part of the settlement; Umbralejo is located in the Sierra de Ayllón in the province of Guadalajara and has a medieval origin like Granadilla. Nevertheless, it was abandoned by its inhabitants in the 1960s and 1970s[26], without external factors that directly caused depopulation. All three were included in the program because they belonged completely to the state. These situations have the advantage of allowing the development and rehabilitation of the whole towns at the same time.

Among the three "abandoned settlements", the town of Granadilla stands out, the only one that has been declared an Historic Site. Due to

---

24 This program is promoted by the Ministry for Ecological Transition and Demographic Challenge and the Ministry of Education, Culture and Sport in collaboration with Education Department of the Castilla-La Mancha, Extremadura and Aragón Governments. Retrieved from https://www.mapama.gob.es/es/ceneam/programas-de-educacion-ambiental/pueblos-abandonados/ (31st July 2018).

25 Currently in force, the last call to participate in the national program "Recovery and Educational Use of Abandoned Settlements" was published in 2018. Ministerio de Educación, Cultura y Deporte, *Boletín Oficial del Estado*, 73, 24th March 2018, 25006–25007.

26 Data and brief overview of the towns on the Abandoned Settlements. Retrieved from https://www.mapama.gob.es/es/ceneam/programas-de-educacion-ambiental/pueblos-abandonados/ (31st July 2018).

its geographical location and/or its heritage, this town has grabbed the attention of visitors and the local people. As a result of this interest, in 2011 the International Centre for Sports Innovation called *El Anillo* was inaugurated: a building parallel to the geography of the town, with a similar circular formality and in a very similar location on a peninsula. Adapted to the unique topography, its program is developed around a circle of 200 m in diameter and 7 m width[27]. This sports centre, created as a response to the lack of activities in the environment, makes it possible to take advantage of the natural environment and the artificial aquatic environment. It brings a set of unique opportunities for the development of a multitude of sports. The increase in sports and leisure activities caused by its construction may, together with Granadilla, be part of the solution to reactivate the reservoir area, respecting the tangible and intangible heritage which exists in Alagón Valley.

Conclusions

The settlement of Granadilla is an exceptional reflection of the capacity of anthropic action. This can drastically and definitively interfere with the appearance of a landscape or the evolution of a town; how man is capable of accelerating the passage of time on heritage to the point of advancing its total ruin and leading it almost to extinction. It is an example of the disturbance of a town's history caused by the resolution of a global, social and territorial problem, such as large reservoir infrastructures.

Granadilla is not only attractive because of its geographical or heritage characteristics, but also because it is a ruin. One of the most striking aspects of the town is undoubtedly its ruinous current state, which gives the place an undoubted romantic halo. From the graphic representations of travelers and romantic artists, the attraction of a ruin has lasted until the twenty-first century. We visit archaeological sites, Roman cities, medieval cities or vestiges of buildings, often driven by their state of deterioration. It is an attraction, the fruit of cultural atavism, which at the same time allows us to exercise our imaginative capacity, overcoming in our intellectual constructions the transcendence of our reference.

---

27 Lozano Bartolozzi and Cano Ramos 2012: 136–143.

No less inspiring is the existence of whole settlements under the waters of the reservoirs. In his novel *Distintas formas de mirar el agua*,[28] Julio Llamazares deals not only with the submerged settlements of Spain. He also, delves into the subject from his point of view, as a native of one of them. In addition to literary references, there are prominent ones from cinematography references. In the case of Granadilla, scenes from the film *Átame* by Pedro Almodóvar were shot in 1990[29] and *La gran familia de Granadilla* in 2013[30], on the management of the state program and the lives of the old neighbors of the town.

The answer to heritage problems arising from the construction of reservoirs has traditionally been the partial or total transfer of those elements considered most relevant for the purpose of their conservation and exhibition. The initiative of the Spanish state in the program for the Recovery of Abandoned Settlements opens up other possibilities of coexistence with the cultural heritage. This assumes a new age, the loss of its previous function, and the appropriation of a new one from the urban and heritage point of view. The orientation of the program seeks to revitalize these settlements to recover the link with the natural environment, necessary in urban times such as the current one. At the same time, it stimulates the conservation of the town and promotes awareness for correct protection, valuation and appreciation of our heritage. Subsequently, sports activities have been added to these actions, mainly attracted by the natural areas and the new reservoir. This heterogeneous cocktail makes possible the recovery of the complete urban structure, once accepted as indispensable, and its necessary cultural redefinition.

Bibliography

Almabro Basch, M. (1971). *El templo de Debod*. Instituto de Estudios Madrileños.

---

28 Llamazares 2015.

29 Rodajes cinematográficos en Granadilla, 2013. Retrieved from http://todopedroalmodovar.blogspot.com/2013/12/localizaciones-granadilla.html (31st July 2018).

30 Documental, "La gran familia de Granadilla." In the TV programm *El Lince 3.0*, 2013. Retrieved from http://www.canalextremadura.es/alacarta/tv/videos/el-lince-30-la-gran-familia-de-granadilla-300613 (31st July 2018).

Alvarado Corrales, E. (2014). "Los barruecos y Granadilla (Cáceres). Paisajes patrimoniales singularizados para el turismo y el desarrollo rural." In Lozano Bartolozzi, M. M. and Méndez Hernán, V. (Eds.), Patrimonio cultural vinculado con el agua: paisaje, urbanismo, arte, ingeniería y turismo, 21–36. Junta de Extremadura, Editora Regional de Extremadura, Universidad de Extremadura and Ministerio de Economía, Industria y Competitividad.

Anonymous. (1963). "Visita del ministro de obras públicas al plan hidroeléctrico y de regadíos de Cáceres." In *ABC newspaper*, 29th January, 47.

Anonymous. (1982). "Inaugurado el aprovechamiento hidroeléctrico de Gabriel y Galán en Cáceres." In *ABC newspaper*, 14th May, 7.

Caballero González, S. (2004). *Granadilla al hilo de la historia: desde los tiempos remotos hasta finales de la Edad Media*. Caja Duero.

Espejel Díez, J. A. (1995). "Un sello concejil. El sello de la villa de Granadilla, en Cáceres." In *Norba: Revista de arte*, 14–15, 7–14.

Espejel Díez, J. A. and Fernández de Castro, C. (1967). *Informe sobre Granadilla*. CHT-MOPU.

Fernández-Daza Álvarez, C. (2005). "Gabriel y Galán o la necesidad de ser en la palabra escrita". In *Alcántara*, 63, 4ª Época, 33–57.

Gutiérrez Macías, V. (1963). "Un albergue de turismo en el castillo de granadilla." In *ABC newspaper*, 31st July, 13.

Junquera, J. and Pérez Pita, E. (2003). *Junquera-Pérez Pita 1969–1999*. Fundación Cultural COAM.

Llamazares, J. (2015). *Distintas formas de mirar el agua*. Alfaguara.

Lozano Bartolozzi, M. M. (1980). "Granadilla: Un conjunto urbano en un medio rural" *Norba. Revista de arte, geografía e historia*, 1, 50–68.

Lozano Bartolozzi, M. M. and Cano Ramos, J. J. (2012). "Buildings that face the water. Mirrors of heritage in the Tajo Basin" In *e-rph*, 10, 113–146.

Maldonado Ramos, L., Vela Cossío, F., and Floriano Martín, C. (1995). *Arquitectura en Granadilla y Umbralejo. Diez años de participación en el programa de recuperación de pueblos abandonados, 1984–1994*. Cursos de Arquitectura de interiores del Departamento de Construcción y Tecnología Arquitectónicas.

Muñoz Hidalgo, D. M. (2010). "Sobre el topónimo Camino de la Plata y el eje S.-N./N.-S. del occidente hispano. Nuevas consideraciones, aportaciones y reflexiones." *El Nuevo Miliario. Boletín sobre Vías romanas, historia de los caminos y otros temas de geografía histórica*, 11, 5–37.

Riaño, P. H. (2017). "La sequía reflota una iglesia románica hundida en el pantano de Campoo." In digital newspaper *El Español*, 30th August 2017. Retrieved from https://www.elespanol.com/cultura/patrimonio/20170830/242976406_0.html (31st July 2018).

Santos, N. (2014). *Lo que el embalse se llevó: pueblos españoles sumergidos.* 10th December, 2014. Retrieved from https://www.traveler.es/naturaleza/articulos/lo-que-el-embalse-se-llevo-pueblos-espanoles-sumergidos-bajo-el-agua/6189 (31st July 2018).

Siliotti, A. (1994). *Egipto: Templos, Hombres y Dioses.* Folio.

Úbeda de Mingo, P. (1985). "Propuesta de rehabilitación en Granadilla, Cáceres". In *Informes de la Construcción*, 36, 369, 5–6.

Section III

Community

A new Phoenix from the ashes?

Marta Marçal Gonçalves, Stefan Rosendahl, and María Teresa Pérez-Cano

1. Introduction and methodology

1.1. Framework

The parish of Cernache do Bonjardim, according to CAOP2011[1], is a small parish in the interior of Portugal, whose history dates back to the Late Bronze Age (about 700 BC)[2]. In 1525, a reference was made to Cernache on a descriptive map that appears in the Codex of Hamburg. The late nineteenth and early twentieth centuries were a time of cultural activity and exceptional dynamism, creating an almost cosmopolitan atmosphere in Cernache do Bonjardim, which is extraordinary, considering the Portuguese conjuncture in those times and the fact that Cernache, then just a village, is located in the center of the country. After this golden age, due to various negative pressures, it gradually declined, until the current stagnation.

The orography of the study area consists of hills, with steep slopes, and agricultural areas, mostly located on plateaus, with small pieces of land and small owners. Cernache currently has 3,000 inhabitants (Census 2011) distributed in a territory with around 7,200 hectares. Desertification and the aging of the population are two of the great problems, being signs of its interiority.

Two opposing forces, water and fire, threatened this territory and its population: in the 1950s, the construction of a dam flooded a large part and, in June 2017, a forest fire devastated about half this territory.

---

1 CAOP means Organizational Administrative Charter of Portugal.

2 Batata 1998.

Despite all this, this community has shown innate resilience, primarily as a result of its persistence and high sense of belonging and pride, improved by the vernacular heritage the people possess, which allows them to remain in this territory, despite the setbacks they faced.

Regardless of the richness of its vernacular heritage, which increases identity, it is an underestimated territory, since all its wealth is little publicized. That's why one of the objectives of the basic thesis[3] was the dissemination, registration, explanation and discussion of its heritage and use, serving as a support for future actions and territorial revitalization.

Those past catastrophic events will make it possible to study whether the population remains resilient, fighting for "their" land and "its" roots, allowing their parish to rise like a phoenix from the ashes.

1.2. Methodology

Bibliographical, iconographic and photographic research was done, simultaneously with field surveys that confirmed the information obtained and which posed new questions to be clarified in the bibliography, constituting a cyclical method of analysis. The bibliography that deals directly with the territory under study is quite scarce, in part due to a fire in the City Hall of Sertã in 1917.

Field surveys are absolutely fundamental to provide direct contact with people and constitute empirical methods[4]. These methods require the observation and investigation of events, verifying how people react to them. In the field under study, the collection of information directly from the territory, the populations and the development of an empirical basis for drawing conclusions constitute an important part of the research.

In order to evaluate the community's resilience, it is necessary to collect qualitative and quantitative data to help to define and understand better the object of study, but mainly to understand the relations between the parties that constitute it and the interactions between the object and its surroundings ("participatory method of action"[5]). For this purpose, direct inquiries were carried out according to the adopted

3 Gonçalves 2016.

4 Martínez Zárate 2006 and Serra 2006.

5 Serra 2006.

method[6], taking the opportunity to collect information directly from the respondents.

The "qualitative method"[7] was applied in order to deepen the knowledge about the study's object, describing the observed events from the approach's point of view.

The evaluation of the community's resilience presupposes the observation and analysis of that community and has an eminently descriptive character about the environment that surrounds the inhabitants, aiming to include the opinion of the people in those situations, and being so-called phenomenological, informative or descriptive surveys[8].

2. Characterization of the Cernache do Bonjardim's parish from the point of view of community resilience

2.1. General considerations. The community resilience

Resilience has been pointed out as a way of predicting and dealing with problems caused by catastrophic or sudden disturbances[9]. It is important, from the perspective of resilience, to make the temporal distinction between a sudden disorder and a *slow-burn*. In the first case we have as examples the tsunamis or the Olympic Games and in the second case, the desertification of the territories or the global climatic change[10]. In both cases, there may be positive and negative disturbances. The territory under study has suffered a kind of slow-burn phenomenon, since the 1960s, at which point it began gradually to lose population[11].

When working on community resilience, bottom-up governance is emphasized, since people in the community need to know what will be best for them.

For this, it is necessary to know each other, which implies an analysis of the negative and positive aspects of the community. For this purpose,

6 Colussi 2000.

7 Martínez Zárate 2006 and Serra 2006 op. cit.

8 Ibid.

9 Atkinson, Dietz, and Neumayer 2007.

10 Pike, Stuart Dawley, and Tomaney 2010.

11 Gonçalves, Perez-Cano, and Rosendahl 2017.

the method proposed by Colussi[12] was adopted, which is aimed at rural communities such as the parish in the study.

In this method it is intended, in a first stage, to elaborate what is called "Community Portrait", CP, which is the "*description of a community from the perspective of resilience*"[13], based on surveys and analysis of several factors. Based on these data, it is possible to elaborate a SWOT analysis, an advanced SWOT analysis and to draw conclusions. This method is based on the so-called "Community Economic Development" Plan, CED, that is, a plan to design a strategic plan for the community, valuing the positive points and improving the negatives. The next phase is the implementation of findings on the ground, but this point is not included in this study.

It is important to note that being more prosperous does not necessarily mean that more money is generated, but rather that greater well-being is generated: as Callaghan[14] says "*the focus should be on* development *rather than* growth". The intention is to increase responsiveness to new ways of thinking and focus efforts on key issues for the community in order to strengthen the community's capacity to lead its own future[15].

From the SWOT analysis[16] it was concluded that: in relation to internal factors, the "strengths" seem to be superior to "weaknesses", with a positive balance. It is a community where everyone tries to involve people in decisions, one in which a general sense of mutual support exists, employment is diversified, there is no fear of embracing another profession, most of the five largest employers in the parish are owned by natural persons and the illiteracy rate is not too worrying. However, there is some confusion as to whether or not there are specific economic objectives for the parish or if there is an economic development plan, CED; education, especially higher education, is not valued. These are aspects whose improvement only depends on the community itself.

With regard to external factors, "opportunities" are apparently inferior to "threats", and therefore have a negative balance. It is an attractive community for those who come from outside; there are many informal

---

12 Colussi 2000.

13 Ibid. pp. 1–21.

14 Callaghan and Colton 2008.

15 Colussi op. cit.

16 Gonçalves 2016.

associations and organizations, many kinds of infrastructures, there are "buy-in" campaigns to which the strong sense of attachment and pride in the community contribute. Nevertheless, it is a community that is threatened by interiority, desertification, population aging, unemployment, the loss of basic infrastructures such as health care, among others. These aspects are exogenous to the community and do not depend only on the characteristics of the community; however, there are some aspects that may cease to be "threats" if some internal work is done. An example of this is the fixing of young people with higher education: young people want to return to their villages, but there are no opportunities, they say. If these opportunities would be created, the likelihood of these youngsters getting fixed increases. This can be done, for example, through incentives to open an office, as in the case of some "business nest" models, or to create possibilities of "home-working".

2.2. *Impact and resilience caused by water*

The characteristics of the Zêzere river before the construction of the Castelo do Bode dam (the *old river*) and after (the *new river*) are quite different. This change had a profound impact on the population of the Cernache do Bonjardim's parish, since, among other things, the best farmlands were flooded (and some compensations were never paid); the systems for grinding the wheat were flooded, so many of them were lost, making it difficult to "*grind the bread*".

The landscape changed: in about two years, a relatively narrow river, with some broad valleys where agriculture was practiced, was changed to an immense mirror of water, with an area of about 32.9 km[17] where the steep slopes of the mountains end. An immense heritage was lost and the landscape was modified radically. The exploration for gold, which was carried out since the Late Bronze Age[18], disappeared in the middle of the twentieth century, leaving only the *conheiras* ("an ancient alluvial exploration of gold, which may have been explored from the Chalcolithic to the twentieth century"[19]) as evidence of this activity.

17 Ambiente n.d.

18 Batata 1998.

19 Ibid.

Fishing was an important element in the medieval economy, and this region was very important due to its richness in fish. Before the Castelo de Bode dam (concluded in 1951), the Zêzere river and its two tributary streams (Cerdeira and Sertã) were abundant in fish, with barbs, bogues, trout, eels, shads and lampreys, for example. With the rise of the waters, most of these species disappeared, so that nowadays only largemouth bass and perch exist[20]. Connected to fishing there are activities such as the manufacture of nets and the construction of the typical boats of the Zêzere river, the *abrangel* or *barco de três tábuas* (three boards' boat), still used today by the riverside populations as a way of mobility.

Much heritage disappeared, as did villages or parts of them: no mills are found today in the village of Moinhos da Ribeira (Stream's Mills); the village of Foz da Cerdeira, like some others was devoured by the floods; the wooden irrigation wheels on the banks of the Zêzere river have disappeared... The construction of the Castelo do Bode dam and its reservoir has had an extraordinary impact on the environment and the socio-economic structure of the region. The filling of the reservoir was a very painful process for this community, strongly agricultural and rooted in the land. The mills and watermills were submerged quickly, and suddenly, and there was nowhere to grind the wheat[21].

The memories of the *old river* persist in some rare photos, in illustrated postcards and in the engravings of Alfredo Keil (a Portuguese composer, painter and poet). The generations born after 1950 have a completely different image, having only known the *new river*, an immense mirror of water with a lush forest landscape. When, in the 1950s, the dam of Castelo do Bode on the Zêzere river was built downstream of the parish under study, there was no concern to mitigate the impacts that could arise. Nevertheless, the Cernachenses (natives of Cernache) reacted, readjusted and reorganized themselves and they survived, demonstrating an innate resilience.

2.3. Impact caused by fire

Nearly every year this region is plagued by forest fires. The question then arises: what was different about the fires in Central Portugal in June

---

20 Gonçalves and Rosendahl 2011.

21 Mendes 2009.

2017? In these fires, about 66 people died (officially 64) and 254 injured, 7 seriously. These fires were considered the deadliest in the history of Portugal[22] and the eleventh deadliest in the world since 1900[23]. Forty-seven of the victims died on the road trying to escape from the fire. These numbers are involved in a controversy, as it is thought that there will be more than 80 fatalities[24].

In Fig. 1 we can see the evolution of the forest fire, seen from space: a) on the eve of the fire, on June 16; b) two days later, on the first day the fire is visible from space; c) the next day, where the huge column of smoke merges with the clouds; and at the end, d) seven days later, showing the burnt area, equivalent to about 53 000 hectares[25].

Several factors contributed to this tragedy: on the one hand the operating system did not work properly, because whenever a fire starts, the local firefighters have to communicate with headquarters in Lisbon and wait for orders, losing any autonomy of command; when firefighters went to help other corporations in distress, such as those of Cernache do Bonjardim, the fire crossed the Zêzere river in areas where the riverbed is narrow, having passed to the parish under study. So, firemen from other corporations needed to come to help because the way back, meanwhile, became impassable. However, these firemen did not know the terrain and did not know where the villages and isolated houses were located (sometimes neither GPS nor satellites can help). Some corporations who fortunately came to help were foreign, but there were no firefighters who knew the terrain to guide them. So, the populations called them and helped them, even though they spoke different languages.

---

22 "Grandes Incêndios Em Portugal. Uma História Que Começou Em 1986," 2017.

23 Barros 2017.

24 Alves 2017.

25 Diário de Notícias 2017.

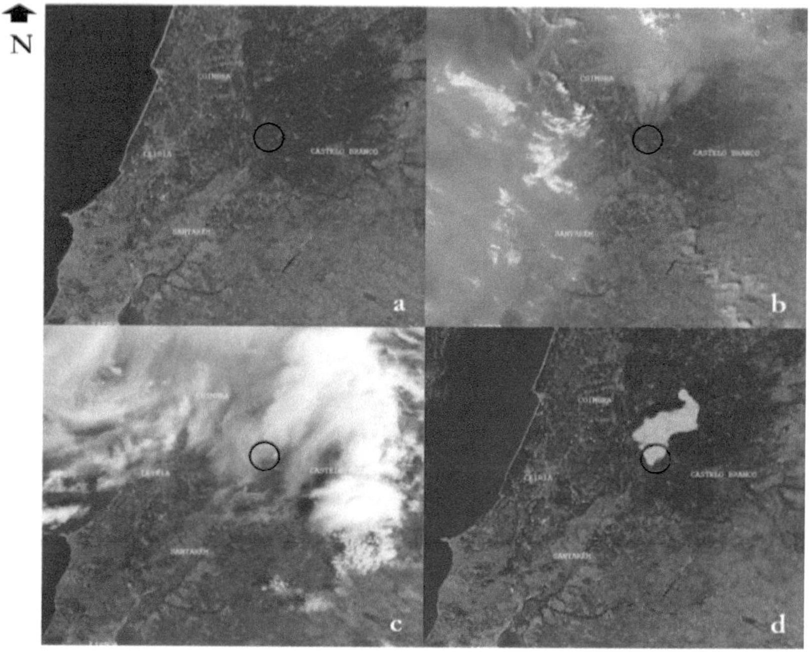

Fig. 1. Forest fire of June 2017 in the Centre of Portugal, with the parish in the study indicated: a) on June, 16; b) on June, 18; c) on June, 19; d) on June 23, light shading: burned area. Source: www.observador.pt, based on NASA images. Accessed on 24/4/2018.

The question of the loss of the decision-making power of the corporation's commanders arose: anyone who knows the area of Cernache do Bonjardim's parish knew that when the fire arrived on the riverside of Figueiró dos Vinhos, in the North, it would pass to the zone of Almegue and Sambado in Cernache's parish, where the river is narrow. It is questioned why this was not considered, knowing the winds that blew that day. At least half of the parish's territory was consumed by fire (Fig. 2, left), but fortunately the damage was only material, and no death occurred there, although the fire was very close to the houses (Fig. 2, right).

The full reliance on satellite images, which in part led to the extinction of the Forest Rangers, has also been called into question, considering presently the reactivation of this unit. A few years ago, watchtowers were scattered across the mountain range, but they were considered obsolete.

Natural factors were also not favorable, because a heatwave with temperatures above 40ºC prevailed, together with a dry thunderstorm[26], according to the official version; on June 18th, there were 163 active fires all over the country[27]. The forest cover, mostly composed of eucalyptus and pine trees, is extremely conducive to the rapid advance of the flames.

Fig. 2. Part of the studied territory (left) and village of Matos do Pampilhal (right), after the June 2017 fire. Photographs taken on July 9, 2017.
Source: Authors.

One of the concerns of the entities was to rescue the elderly because, besides being the majority, they have reduced mobility. However, many young and old people did not want to leave their homes, preferring to stay and defend them, even if it would cost them their lives. This reaction, which may be considered unconscious by many, is not strange to those who know the people of this part of Portugal, so linked to their land, with a very strong sense of belonging and identity.

Leaving their home and not being able to defend their belongings, means to tear the roots that bind them to their land, as if they were lost in a strange world. To see the work of the whole year and life destroyed, to hear the sound of the fire advancing, to smell the smoke, the forest's ashes and the burnt animals, are experiences that take a lot of time to forget. The struggle of man and fire filled the days: when there was water

[26] Naves 2017.

[27] SAPO 2017.

in the wells, the electric pumps did not work because the cables had been incinerated (Fig. 3, left); when you could get water, the hoses were burned… meanwhile, the wind changed direction again, burning what was left… the feeling of impotence is enormous (Fig. 3, right). Gigantic disappointment and frustration. There was very little help. The urbanite, who has never experienced this feeling of union with the environment felt in the rural world, has difficulty understanding the love for the animals, trees and land that the farmer has.

Fig. 3. Melted electric cables (left) and Casal do Moinho (right), after the June 2017 fire. Photographs taken on July 9, 2017. Source: Authors.

In the parish of Cernache do Bonjardim, losing belongings means losing the traditional and still used form of subsistence agriculture: people, when they leave their jobs, go home to work on the land they own and feed domestic animals. That is the reason why it is almost impossible to sell houses without some land annexed, the *leiras*.

It is early August 2018 and most of the millions of euros gathered due to the solidarity of all Portuguese citizens, to help those injured by these fires, have not yet reached the destination and some believe they will never arrive. However, people stayed, fought, helped each other and tried, as much as possible, to reorganize their lives, having proved and proving that they are resilient. More elements will be gained only over time; in this article we can only report what is happening a year later. At the next Census we will have more to count on this subject.

3. The future

Based on the SWOT analysis carried out by Gonçalves[28], an advanced SWOT analysis was performed, making the interception between the various fields of this analysis (Tab. 1), and not only between endogenous and exogenous factors.

Tab. 1. Advanced analysis of the various items of a SWOT analysis. Authors' elaboration.

<table>
<tr><th></th><th></th><th colspan="2">External factors</th></tr>
<tr><th></th><th></th><th>Opportunities</th><th>Threats</th></tr>
<tr><td rowspan="2">Internal factors</td><td>Strengths</td><td><u>Leverages</u>
They are forces that are or may be linked to opportunities.</td><td><u>Vulnerabilities</u>
These are the forces that meet with non-controllable threats; they are factors that can be adverse in the short, medium and long term and need to be filed with defence strategies.</td></tr>
<tr><td>Weaknesses</td><td><u>Constraints</u>
They are the result of the relationship between weaknesses and opportunities; that is, what cannot be exploited due to weaknesses. These weaknesses need to be eliminated to conquer opportunities.</td><td><u>Problems</u>
This field constitutes the real problem; they are the points that intersect in weaknesses and threats.</td></tr>
</table>

This analysis consists[29] of the attribution of weights to each factor listed in a range between 100 (most important) and 0, based on the probable impact of this characteristic on the strategic position in the community. These weights must add up to 100 in total, for both internal and external factors. In the same way, a factor of importance, between 1 (weak) and 3 (excellent), is attributed to each characteristic, based on

28 Gonçalves 2016.

29 Aprenda a Análise SWOT Com Modelo No EXCEL n.d.

the community's response to this factor. The result for each characteristic is obtained by multiplying the weight and the importance factor. The ranking of importance in the advanced SWOT analysis is obtained by the sum of this result for each interception carried out.

As for the method adopted to define the community's resilient portrait (CP), certain relationships between some characteristics exist. This allows us to reduce the number of interceptions and to carry out only those that are related. The following conclusions were drawn from this interception:

- The three main points of leverage are the attractiveness of the community to outsiders, the diversity of employment and the fact that 60 % of the five largest employers are natives of the parish. The main characteristics that attract outsiders are the optimism of the community, the faith that they will be able to overcome this bad period, the high level of pride shown and the existing "can do" attitude, believing that they are able to influence the future by working hard on issues that directly affect them. Subsequently, as leverages, there are: the existence of a large number of informal groups, with good cooperation between them; public participation that is encouraged and solicited; a concern to have compulsory schooling; and the improvement of accessibility, which facilitates out-of-community and professional contacts that can boost the economy.
- In spite of the economic situation and the optimism that exists in the community, there is a risk that all of this is compromised by the fact that there are few children under the age of 14, indicating a decrease in the number of young families and the aging of the population. The feeling of attachment is very high, as is the level of pride, so that despite the difficult economic environment, unemployment and poverty, most people continue to stay and fight for their community, cooperating with one another. This mutual support and the positive reception of new ideas in doing things, is threatened by the aging of the population, and the inability to see continuity or a future. In this community there is also a "reinforcement" of the elderly population by the settlement of retired couples from outside the parish. All this put at risk the existing optimism and sense of pride, which, despite being felt in the younger ones, will tend to disappear. Gradual depopulation and loss of heritage is also a risk. The non-valuation of the property belonging to natives,

and the absence of a study that makes it possible to identify and help the companies that are in danger of closing, does not encourage the creation of new ones. The lack of a general economic plan for the community (CED) with a common goal hinders cooperation between organizations, as well as changes and more active participation of the population in matters that concern them directly. Traditional education is not valued, partly because there is no employment for young people with a higher degree, which makes it difficult to change the way of thinking. These were the main vulnerabilities encountered.

- As main constraints, a large number of informal groups, organizations and associations have been set up, but meetings on community affairs are sparsely attended, with people not taking the opportunity to express their opinion, and there is no clear perception whether or not public participation influences the most important decisions for the community. This does not facilitate the change of mentalities that the economic crisis requires. When there are issues involving neighbors, most prefer not to act; nevertheless, the parish has good external attractiveness. The crisis has also brought back people of working age who have lost their jobs in large urban centers, but are usually not skilled. This trend can be countered by the fact that accessibility has been improved, and advantage taken of this to generate some market opportunities and the creation of new products.
- The major problems encountered are unemployment, aging populations, lack of a common economic plan (CED), no definition of strategies for the community, with the stagnation of mentalities: most people still think that a large employer would be the best solution to unemployment. The question of not identifying and helping companies that are in danger of closing does not make it possible to direct resources, causing a failure to define commercial areas with more or less success in the conjuncture of the parish, thus making investment in education itself attractive in order to respond to requests that would appear.

This analysis confirmed in general the conclusions of the SWOT analysis presented. The leverages (about 23) are almost as many as the problems counted (about 25), both below the vulnerabilities (about 49) and above the restrictions (about 13).

The parish has many vulnerabilities because, despite its strengths, it is very threatened by external factors; these vulnerabilities may improve while leverage almost equals problems; apparently, the community is taking advantage of some of the opportunities that arise, despite their weaknesses (constraints).

4. Some final remarks

Throughout the history of the Cernache do Bonjardim's parish, there have been moments of great impact on the population; curiously, the largest ones, are due to two opposing elements: water and fire. The first with the construction of the Castelo do Bode's dam with its reservoir, which has submerged an immense heritage and also the best farmland, changing the physiognomy of the landscape. At that time, the population adapted, reorganized and survived. The second was with the catastrophic fires of June 2017. These forest fires were especially prolonged and burned an immense area of land, not only in the parish, but practically the whole center of the country.

Due to the analyses presented here, having as main strength the sense of attachment and pride in their land, there is a real hope that the population will adapt, reorganize and survive. In fact, there was apparently no abandonment of the houses or the land, and in 2018, a year later, the fields have been planted again, there are vegetables and fruits and the domestic animals are seen again.

As a final question, will we have "a new Phoenix from the ashes?" From the analyses made and the knowledge acquired on the ground with the population, we can almost certainly say yes! Cernache do Bonjardim will be reborn from the ashes. However, this will only be confirmed in the coming years, and with the upcoming Census.

The hope of renewal and rebirth, after such a tragedy, is real; and from the ashes, a flower is born (Fig. 4). It is important to see these tragedies not as threats, but as opportunities, making the best of misfortune.

Fig. 4. The parish of Cernache do Bonjardim, a shoot of a plant, after the fire of June 2017. Photograph taken on July 9, 2017. Source: Authors.

Bibliography

Alves, A. A. (2017). Quantas pessoas morreram na tragédia de Pedrogão Grande? Retrieved July 29, 2018, from O Insurgente website: https://oinsurgente.org/2017/07/22/quantas-pessoas-morreram-na-tragedia-de-pedrogao-grande/.

Ambiente, A. P. do. (n.d.). Barragens de Portugal. Barragem de Castelo do Bode. Retrieved July 27, 2018, from Agência Portuguesa do Ambiente website: http://cnpgb.apambiente.pt/gr_barragens/gbportugal/CastelodoBode.htm.

Aprenda a análise SWOT com modelo no EXCEL. (n.d.). Retrieved March 1, 2016, from http://scientiaarca.com.br/analise-swot-excel/.

Atkinson, G., Dietz, S., & Neumayer, E. (2007). Introduction. In G. Atkinson, S. Dietz, & E. Neumayer (Eds.), *Handbook of Sustainable Development* (pp. 1–23). Cheltenham: Edward Elgar Publishing Limited.

Barros, R. (2017). Fogo de Pedrógão Grande é o 11.º mais mortal do mundo desde 1900. Retrieved from Rádio Renascença website: http://rr.sapo.pt/especial/86656/fogo_de_pedrogao_grande_e_o_11_mais_mortal_do_mundo_desde_1900.

Batata, C. (1998). *Carta arqueológica do Concelho da Sertã*. Sertã: Câmara Municipal da Sertã.

Callaghan, E.-G., & Colton, J. (2008). Building sustainable & resilient communities: a balancing of community capital. *Environment, Development and Sustainability*, *10*, 931–942. https://doi.org/10.1007/s10668-007-9093-4.

Colussi, M. (2000). *The Community Resilience Manual. A Resource for Rural Recoverty and Renewal*. Port Alverni: Centre for Community Enterprise.

Diário de Notícias. (2017, June 23). Incêndios: Quase 53 mil hectares arderam na região centro. Dados provisórios. *Diário de Notícias*. Retrieved from https://www.dn.pt/lusa/interior/incendios-quase-53-mil-hectares-arderam-na-regiao-centro---dados-provisorios-8586699.html.

Gonçalves, M. M., & Rosendahl, S. (2011). Impacte das grandes obras de engenharia. A barragem do Castelo do Bode e a freguesia de Cernache do Bonjardim. *ICEUBI2011 International Conference on Engineering UBI2011 Innovation & Development*, Paper 39. Covilhã: Edições UBI 2011.

Gonçalves, M. M. (2016). *El conocimiento del patrimonio en cuanto factor de estructuración de la sostenibilidad de los territorios. El caso de la freguesia de Cernache do Bonjardim, Portugal* (ETSA Sevilla). Retrieved from https://www.educacion.gob.es/teseo/mostrarSeleccion.do.

Gonçalves, M. M., Perez-Cano, M. T., & Rosendahl, S. (2017). Heritage as a structuring factor of the Territories' resilience. In *Asia Pacific Handbook of Contemporary Research. Volume 1* (1st ed., pp. 44–60). Retrieved from http://apiar.org.au/?page_id=1999.

Grandes incêndios em Portugal. Uma história que começou em 1986. (2017). Retrieved July 27, 2018, from Rádio Renascença website: http://rr.sapo.pt/noticia/86522/grandes_incendios_em_portugal_uma_historia_que_comecou_em_1986.

Martínez Zárate, R. G. (2006). *Manual de Tesis. Metodología especial de investigación aplicada a trabajos terminales en arquitectura*.

Mendes, E. (2009). *Dornes, o tesouro dos Templários*. Dornes: Junta de Freguesia de Dornes.

Naves, F. (2017, June 18). Mistura explosiva: Trovoada seca, calor e vegetação densa. *Diário de Notícias*. Retrieved from https://www.dn.pt/sociedade/interior/mistura-explosiva-trovoada-seca-calor-e-vegetacao-densa-8572702.html.

Pike, A., Dawley, S., & Tomaney, J. (2010). Resilience, adaptation and adaptability. *Cambridge Journal of Regions, Economy and Society*, *3*, 59–70.

SAPO. (2017). Proteção Civil: Saiba quais os incêndios ativos em Portugal. Retrieved August 4, 2018, from pplware, based on www.prociv.pt/ website: https://pplware.sapo.pt/informacao/protecao-civil-saiba-quais-os-incendios-ativos-portugal/.

Serra, G. G. (2006). *Pesquisa em arquitectura e urbanismo: um guia prático para o trabalho de pesquisadores em pós-graduação* (1st ed.). São Paulo: Editora Mandarim e EDUSP.

Building community resilience through heritage in small towns: Case study of Vysoké nad Jizerou

Jiří Janáč

This exploratory chapter seeks to evaluate and test the usefulness of the concept of resilience understood through the lens of cultural heritage as an instrument for building more resilient small-town communities. Small towns, with their specific character resembling a community (one of the many definitions of a "small towns" in existing literature draws on this distinction:)[1] provide a specific cultural and social context. The hypothesis promoted and tested (to a limited extent) in this chapter proposes that heritage activities (understood in broad terms) are central for building of more resilient communities in small towns and peripheral areas facing marginalization in the ongoing process of globalization. Cultural heritage production, that is, mobilization, performance and reproduction of traditional local identities (incl. heritagization) represents a viable strategy for former micro-regional economic centres to survive and recover from adverse conditions and retain their urban character. On the other hand, strategies of heritage mobilization remained constrained by the self-image of the community and its collective memory.

The chapter opens with a brief theoretical discussion of crucial concepts and then moves to the case study of the small Czech rural mountain town of *Vysoké nad Jizerou* (Vysoké on Jizera river), and discusses the role of heritage mobilization in tackling the challenges of post-socialist transformation.

1 Steinführer et al. 2016 and Klusáková et al. 2017.

Theoretical considerations: Resilience and heritage in small towns

"Urban resilience" has been formulated from the planning perspective as the ability of the urban system (consisting of both material and human dimensions) to survive various types of risks and hazards, both natural and man-made[2]. In such a form the concept of resilience entered heritage studies following the last economic recession. Building on the (often implicit) premise that the cultural heritage embedded in the local environment (in our case a small town) provides the symbolic basis for the identity of people and place, heritage experts argued that the urban system cannot recover the integrity and character of the place after a disaster without the restoration/survival of the local cultural heritage. Thus, by shifting attention towards the human part of urban system, emphasizing the role of heritage as a major prerequisite for community resilience[3], they called for the inclusion of heritage as an indispensable prerequisite for any successful attempt at post-disaster reconstruction as a part of disaster risk reduction (e.g. various UNESCO initiatives)[4].

Unfortunately, the planners' perspective excludes agency and dynamic from the heritage production process; it focuses on "shocks" and obscures long term developments and continuities[5]. We believe that "heritage resilience" discourse needs to address the more clearly assumed cultural dimension of resilience. Following a critical turn in heritage studies, we understand cultural heritage broadly as the legacy (both tangible and intangible) inherited from the past generations, which at a certain point acquires symbolic value for a given community. Instead of being an objective analytical category, we understand cultural heritage as an actor category, subjected to constant re-evaluation and re-definition by the community. In effect, heritage is permanently disputed and re-negotiated, as the community responds and adapts to external pressures – "saving heritage is nothing but a specific manifestation of change"[6]. Literature focusing on cultural practices of heritage

2 Meerow et al. 2016 and Zhang and Li 2018.

3 Holtorf 2018.

4 UNESCO, NICH 2016.

5 Bollig 2014.

6 Holtorf op.cit.: 647.

production as a means of building more resilient communities remains so far relatively limited[7].

Methodologically, our research builds upon anthropologic methods of heritage studies – we have conducted a set of interviews with leading figures of the heritage debate in Vysoké nad Jizerou. However, we also drew inspiration from memory studies, especially concerning the formation of the town's identity[8]. Besides heritage literature, local histories and various materials published by local stakeholders, we have conducted a series of interviews with citizens of Vysoké.

The town of Vysoké nad Jizerou

Contemporary Vysoké nad Jizerou is a rural small town par excellence. With the current population of less than 1000 inhabitants in the urban core (with associated villages the number rises to 1300), it ranks among the smallest municipalities in the Czech Republic[9]. From the perspective of a visitor, the town preserved its traditional rural character, having missed both waves of industrialization of the region (the late nineteenth century boom in textile manufacturing and the so-called socialist industrialization of 1950s). The persisting rurality of Vysoké, at least in the eyes of the observer (tourist, visitor) is further underlined by the land-use pattern of the town's territory, dominated by agricultural lands and forests on steep hillsides (out of 2066 hectares built-up areas cover only 26, or 1.3 %, while forests cover 32 % and 60 % cropland and pastures)[10].

At the same time, the town of Vysoké features some of the typical signs of urbanity. First, the material urban structure, and not only the built environment, clearly fits the traditional municipal set-up developed in central Europe since the Middle Ages – a radial street system with central market square in the middle, a town hall, a municipal park (orchards) and a theater building (Fig. 1). Vysoké lacks another standard component of a historic town – the residence of the local nobility – but the lavishly designed summer residence of the first Czechoslovak prime

7 See for instance Beel 2017.

8 Halbwachs 2008.

9 Vaishar et al. 2016.

10 Vyhodnocení vlivů 2015.

minister and local native Karel Kramář (in office 1919–1920) compensates (at least visually) for such an imperfection (current owners even call it "Kramář castle": http://kramaruvzamek.cz/).

Fig. 1. The town of Vysoké nad Jizerou (Hochstadt an der Iser) on the Imperial Obligatory Imprints of the Stable Cadastre developed between 1826–1843. Source: archivnimapy.cuzk.cz.

Second, Vysoké retains some elements of centrality within a regional settlement system, serving functions considered to be the defining characteristics of a "town" in social geography[11]. While it has been gradually stripped of its central administrative role since the mid-nineteenth century (it used to be the centre of a judiciary district until 1949) and lost some other locally important social functions (the general hospital shut down in 1956), the town boasts a fully developed educational system (an elementary school and a high school) and provides various other services for the wider area: post office, petrol station, supermarket, state administration (register) etc[12].

Furthermore, since the emergence of modern (mass-)tourism in Central Europe in the late nineteenth century, Vysoké, due to its position in the foothills of the Krkonoše (Riesengebirge, Giant Mountains),

[11] Steinführer et al. op.cit.

[12] Lukáš 1992.

assumed the character of a tourist resort town, initially popular among early Czech adherents of mountaineering and advertized as climatic spa (clean mountain air). Over the years, winter tourism transformed into one of the principal (if not the principal) economic sectors driving and deciding the wellbeing of the town, contributing significantly to its "urban" character by fueling the development of the related service sector[13].

Construction of local identity heritage prior to 1990s

The major features of the town identity developed in the second half of the nineteenth century in the context of the nation-building movement – a development Vysoké shared with majority of historical towns in Central Europe[14]. The town located at the foothills of the Giant Mountains (Krkonoše), on the top of a hill at an altitude of 700 m above sea level, gradually developed an image of an outpost of ethnic Czech settlement and culture – both in the vertical sense (located high in the mountains) and the horizontal sense (being the centre *of a Czech language region encircled by a German-speaking majority – see* Fig. 2).

[13] Vysoké nad Jizerou 2015.

[14] Ira 2013.

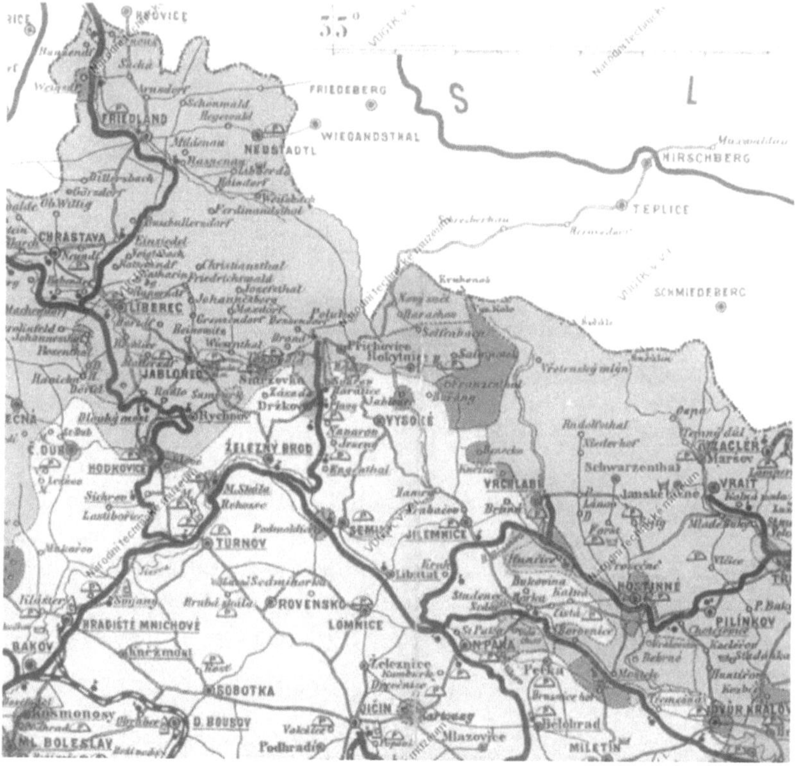

Fig. 2. Section of the 19th century ethnographic map of Bohemia based on the 1880 population census (Czech-light grey; German – medium grey; mixed – dark grey) showing position of Vysoké nad Jizerou encircled by German speaking areas. Source: Národnostní mapa Království Českého na základě sčítání lidu z roku 1880.

The distinctly Czech character of the town and adjacent villages, unique in the Krkonoše region otherwise dominated by a German-speaking majority[15], predestined the town to become a natural centre of the Czech national movement and culture in the nineteenth century. The "outpost" position of the Vysoké "Czech enclave", perhaps together with long, dark and snowy winters, inspired the organization of various nationally framed cultural activities very early on. Since 1789, Vysoké has

15 Klapka and Martinát 2005.

had a Czech-language amateur theater, one of the first in the country[16]. The local population manifested its strong national feelings in meetings and gatherings held on the nearby hills in the 1860s and early 1870s and continued to define itself as a stronghold of the Czech nation at least until the transfer of the German population away from the Krkonoše Mountains after the Second World War[17].

Prague based intellectuals and leaders of the national movement in the late nineteenth century admired the specific local culture and dialect of the Vysoké region and interpreted it within the frames of the national myth based on the contemporary Herderian image stressing the rural roots of the Czech nation[18]. The people of Vysoké and the surrounding area quickly embraced the idealized image of the Czech Highlander ("*horák*"), a poor and slightly coarse character, who is however unyielding, tenacious, pious (spiritual), uneducated but wise, and extremely loyal to the nation. The identity of the "Czech Highlanders" by definition incorporated a strong sense of locality – distant from national cultural centres and defined by harsh physical environment, romantically depicted in an influential country-prose novel "Forgotten Patriots" published in 1894 in national newspapers. In the book, vaguely based on the diaries of a local teacher and violin maker, Vysoké was renamed Větrov (Windtown) to emphasize the inhospitable climate conditions[19]. The nickname stuck and continued to be used on various occasions by locals up until now (Fig. 3). The Czech & Slavonic Ethnographic Exhibition, which took place in Prague just a year later, provided an opportunity for the expression of such a local identity – and the town of Vysoké prepared a representative and lavish installation depicting the hard, but civilized and honorable life in the mountains, with its specific features such as home-made mechanical Nativity Scene Sets and local customs and mythology[20].

16 Klíma 1993: 233.

17 Skrbek 1936.

18 Macura 2015: 435–454.

19 Rais 1894.

20 Dědovský et al. 2019.

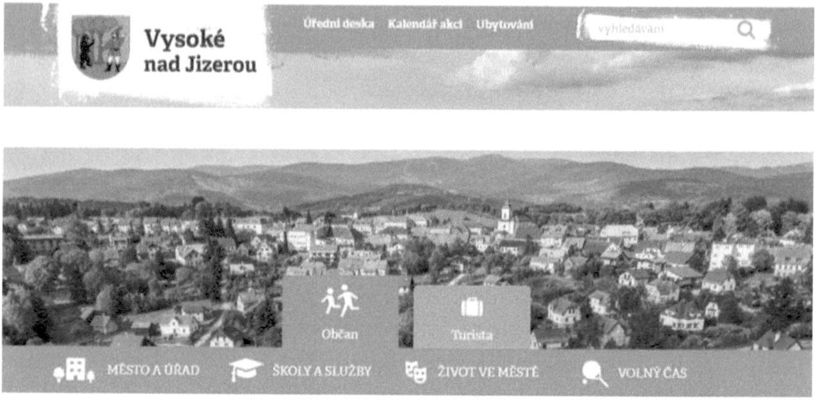

Fig. 3. The current website of the municipality does indeed visually enhance the close connection between the town and Krkonoše mountains. Source: https://www.vysokenadjizerou.cz/.

The image of Vysoké as a cultural centre of ethnically Czech Highlanders of the Krkonoše Mountains resulted in successful "place-making" activities symbolized above all in three projects designed to confirm the urban and cultural character of the town between the 1880s and the 1930s. First, 1885 witnessed the establishment of the "public park", an almost unique feature in towns of such size. The park, significantly extended in 1895, offered a nice view of Krkonoše Mountains and entangled them symbolically more closely into the town environment. The importance given to theatrical tradition materialized in the erection of a theater building next to the church in 1920s. This tradition has never been lost, the local theater society Krakonoš is still in operation, and since 1971 a nation-wide festival of "Czech rural amateur theatres" takes place in Vysoké every autumn. Third, local identity institutionalized in the form of Vysoké Regional Museum, established by the municipality in 1930. Its collections and installations initially consisted of the remnants of the presentation and collections prepared for the 1895 national ethnographic exhibition, and emphasized and celebrated rich cultural life of "Highlanders" and "forgotten patriots", their customs and traditions, contrasted with the harsh living conditions as reflected for instance in the extremely modest local dietary culture, based partly on local plant varieties (*vysocké zelí* – local variety of cabbage) and regional cuisine (*sejkory* – potato pancakes, *kyselo* – bread soup etc.)[21].

The mountains played a major part in another and more recent layer of Vysoké identity. The so-called nationalization of space, a result of the intersection of development of (mass)tourism and nation-building, resulted in the formation of two competing tourist infrastructures in the Krkonoše Mountains – one ethnically Czech and the other German[22]. In the context of the emergence of early tourism and mountaineering, Vysoké (together with Jilemnice) became a gate to the mountains for Czech tourists, despite the fact that ethnically German towns such as Vrchlabí, Špindlerův Mlýn or Harrachov enjoyed a more favorable location[23]. The establishment of two local branches of the Club of Czech Tourists (est. 1888) in Krkonoše in Jilemnice (the second regional branch altogether – 1889) and Vysoké (1901) testified to their position in this respect[24]. During the 1920s and 1930s the town hosted a number of figures belonging to Czech elites, which did not come only to enjoy fresh mountain air, but also to support the hard-living Czech Highlanders[25]. By early 1900s Vysoké also became associated with winter sports, as one of the first Czech ski clubs was established here and the first long distance race and later national championships in Nordic skiing took place here[26]. Again, the second club only to nearby Jilemnice and Prague, but preceded by almost a decade by German ski clubs in the region – Špindlerův Mlýn 1890 and German Krkonoše Ski Club 1896[27].

Already by the early twentieth century tourism started to significantly alter the urban space of Vysoké -besides small hotels and inns, a pavilion was built on the main square, which featured a plastic model of Krkonoše Mountains. A dormitory of the Ski Club was erected in 1913, and was followed in the 1920s by two new hotel buildings and the municipal hotel Vysocká bouda (1937). The support of the municipal authorities for tourism met with the interests of residents and countrymen living in

---

21 Ibid.

22 Bartoš 2000.

23 Štemberk 2013.

24 Brynda, Dostál, and Pater (1938): 170—172.

25 Vařeka and Plessingerová 1991; Jakl 2004.

26 Lukáš op.cit.

27 Lenderová et al. 2013: 385.

Prague, who launched together a campaign in the late 1930s promoting Vysoké as a winter sports resort. Since then, Vysoké has used the nickname "town of the most beautiful winter" (Fig. 4)[28].

Fig. 4. The town of the most beautiful winter. Postcard from 1940s. Source: Author's private collection.

Over the first half of the twentieth century the national movement and the expansion of mass tourism transformed both the built environment of the town and the identity of its residents. The town was transformed into a modest mountain resort with a reputation of a holiday town of the Czech national intelligentsia[29]. In the process, the local community actively participated in the invention of its identity and started to construct, protect and perform its own heritage based on the (invented) traditions of the (imagined) community of Czech Highlanders of the Krkonoše Mountains. To this end, besides the theater and the museum, local branches of the Czech Ski Club and the Czech Tourist Club also acted as influential memory institutions, especially after the Second World War. Before the 1950s, a significant part in heritage production

[28] Vodseďálek et al. 2004: 65–79.

[29] Jakl op.cit.

had been played also by another group – The *Spolek rodáků a přátel Vysokého nad Jizerou a okolí v Praze* (Union of compatriots and friends of Vysoké nad Jizerou in Prague), established in 1937[30].

These quasi memory institutions in practice resembled more community heritage organizations, as they combined heritage interests with other community services and depended to a large extent on voluntary work. In effect, Vysoké developed during the first half of the twentieth century into a "heritage community," with a strong place identity, clearly articulated concept of heritage, and functioning institutional infrastructure.

Neither the Nazi occupation nor the Communist takeover of 1948 and subsequent social transformations (collectivization of agriculture) left a permanent mark on the local identity. The national heritage policy of state socialism, influenced by Communist ideology, held folklore and social history in high regard, and further strengthened the popular "*horák*" identity, while supressing its religious aspects and remembrance of some of the local great figures (such as Karel Kramář) who did not fit the ideological demands[31]. By the 1950s several houses in Vysoké, characterized as typical examples of local vernacular architecture, were listed as cultural monuments and protected under the national heritage law, together with examples of local art and traditions. The activities of the museum, as well as the theater, enjoyed stable support from the national and local authorities, and representatives of both institutions refer especially to the 1980s as "golden times" (information gained from interviews).

Combined shocks of 1990s

The notion of resilience in urban studies discourse revolves around the capacity of urban systems, societies and communities to absorb and survive external shocks. While in the case of Vysoké we cannot identify a recent natural catastrophe or similar phenomenon endangering physical survival and future existence of the town, following existing literature we define "shock" in less straightforward manner as a "sudden and external" change. Following the collapse of the socialist economy, Vysoké faced a

30 Vodseďálek et al. op.cit.

31 Štulc 2012.

series of pressures which, in their combined effect, could be reasonably described in terms of a "shock". First, the economic transformation and its impact on the economy (tourism, agriculture etc.) undermined economic and social stability; second, climate change brought a wave of weak and less snowy winters, which exerted further pressure on the local tourism industry; third, small towns and rural communities experienced a speeded-up form of post-socialist shrinking – the revival of capitalist economy under neoliberal governments exposed small towns to pressures associated with globalization and concentration of population and social activities on the higher geographical level[32].

Following the collapse of socialism, Vysoké faced difficult challenges associated with a transition to market economy, globalization and "climate change". Small town research in the post-socialist context usually focuses on the economic transition and basically fits in the narrative of urban decline, stressing the issues of de-industrialization, de-urbanization, negative demographic trends and infrastructural shrinkage[33]. For the Czech Republic, such developments have been confirmed by Zemanová for the period of 1990s[34]. Vysoké did not escape a similar fate. Gradually, it lost its character of a micro-regional market centre due to the rapid development of individual mobility- perhaps best illustrated by the emergence of hypermarkets in more distant and larger centres[35]. In spite of such a negative economic situation, Vysoké did not lose a significant part of its population (see Tab. 1).

Tab. 1. Table showing development of the Vysoké nad Jizerou population and number of houses since mid-19th century – *indicates changes in the statistics methodology. Source: Czech Statistical Office (2015).

Year	*1869*	*1890*	*1910*	*1930*	*1950*	*1970*	*1991*	*2011*
Population	3838	3993	3661	3099	2328*	1676	1417	1292
Houses	659	626	630	646	680	459*	649	706

32 Vaishar et al. 2015.

33 Ibid.: 1396.

34 Zemanová 2012.

35 Vaishar 2004: 312.

Restitutions of properties nationalized after 1948 and privatization of land after 1989 followed the pattern common for the entire country and did not lead to revival of a small-scale private farming, characteristic for a rural town prior to the socialist collectivization. Most landowners rented their newly acquired lands back to the successor of the former collective farm[36]. In effect, employment in agricultural sector remained relatively low and stable and the majority of the workforce continued to work in the service industry, most notably tourism.

The opening of borders, low transportation costs and inadequately developed consumer culture (reservations towards the idea of commodified authenticity and ensuing disregard for rural tourism among Czechs) in the 1990s resulted in a significant drop in the numbers of tourists visiting the Czech countryside[37]. While there are no specific data available for Vysoké, general data on the development of tourism in the Czech Republic in 1990s show clearly a significant drop in interest, perhaps most visible in rural areas[38]. Unclear property rights, privatization and the shock therapy economic doctrine decline (56 % inflation) together with the transition from the mass-organized "collective" tourism characteristic for the socialist period further exacerbated the negative trend.

As of 1990, tourist infrastructure in Vysoké consisted of several large-scale buildings owned and operated by industrial companies as their recreational facilities. According to a census of 1991, by 1990 the town of Vysoké, with ca 1000 inhabitants (not including those living in associated villages) offered capacities of about 500 beds[39]. However, the 1990s witnessed a shift towards accommodation in private rooms (from 160 to 400 by 1992), fueled by the short-lived influx of German tourists. By the mid-1990s, most of the capacities were privatized and struggling to survive or abandoned. The former summer residence of Karel Kramář, his villa, represents the most illustrative example of the negative impact of economic transformation on local tourism. Refurbished in the late 1980s by the National Council of Trade Unions (ROH) to fit the needs of socialist workers, the villa irretrievably lost its lavishly designed interiors

36 Swain 1999.

37 Hájek 2002: 561.

38 Vystoupil et al. 2010: 96.

39 Lukáš op.cit.: 84–85.

and has struggled ever since to survive as a hotel[40]. The impact of the transformation period also contributed to problems with the ownership of the tourist/municipal heritage infrastructure. This is best documented by the fate of the local outdoor swimming pool, built in the 1920s. After restitution in the early 1990s the new owner closed the premises and since then has been negotiating conditions of renovation and reopening with the municipal authorities. In the meantime, the site deteriorated to the level of virtual disappearance (wooden structures)[41].

Climate fluctuations further contributed to paint the dim picture of a town abandoned by tourists even darker. The winters of the 1990s have been characterized by an extreme lack of snow[42]. Further, Vysoké could not compete with more favorably located ski resorts higher in the mountains (more snow, better slopes) and started to lag behind significantly in infrastructure investments (artificial snow, chairlifts, higher quality accommodation etc – the process started already in 1970s).

Strategies of heritagization as a form of resilience

The late 1980s were characterized by a highly technocratic approach to heritage management. Expert-led evaluation of the heritage resulted in a focus on the protection and preservation of specific events and objects, but largely ruled out the participation of the local population in the decision making and defining of the heritage value. Community heritage, if not in accord with official national heritage policy, was not granted protection. Furthermore, such an approach created a clear dichotomy between formally acknowledged heritage sites and the rest[43].

In case of urban heritage, from the 1950s state institutions focused on the preservation of historic towns, but Vysoké did not make it onto the list. Its status of "unprotected" town (vastly rebuilt after the 1834 fire) then resulted in a major blow to the local built heritage in the late 1980s, when two blocks of houses on the northern side of the square were demolished. It was a story Vysoké shared with many other (even some of the historical and thus protected) towns in the period when heritage conservation was

---

40 Průková 2007.

41 Větrník 2017: 3.

42 Němec and Zusková 2005.

43 Kibic and Vošahlík 2012 and Krejčí 2012.

governed by the principle of "tolerance to change"[44]. A new Town Hall, opened in 1993, replaced the historically significant Inn "U landvéru," a centre of social life of the town since the mid-nineteenth century[45].

After 1989, liberalization of heritage management marked a significant shift towards the heritagization of the square, aiming at the rectification of past (socialist) grievances[46]. After prolonged discussions, a restoration of the Marian column demolished in 1957 took place in 2000[47] and the square was renamed after Karel Kramář (a figure marginalized during state-socialism), whose bust was placed on his birthplace in the corner of the square in 2016.

The liberalization of heritage management, however, also led to a significant drop in financial and institutional support from the state and regional authorities. The transformation period brought about significant redrawing of the heritage management landscape in Czech Republic with potentially fatal impact on local heritage institutions. In case of Vysoké, such an effect was combined with a tourism crisis.

However, the strong local identity and cultural capital gathered over previous decades made the Vysoké heritage particularly resilient. By the end of the 1980s, almost half of the population of the town engaged in the activities of the ski club and tourist club[48]. The amateur theatre, which was transformed into an NGO, Větrov, in the early 1990s, had around 120 actively participating individuals – most from Vysoké (information gained from interviews). Most of these people proved to be true patriots and did not leave the town during the crisis. Furthermore, ensuing developments clearly demonstrated that Vysoké formed a strong and single "heritage community," which proved to be capable of acting in relative consensus, based on the generally accepted image of Vysoké heritage.

Since 1989, the Vysoké community mobilized the heritage on various occasions and such heritage activities (performative or protective) significantly contributed to the survival of the "urban" and central character

44 Štulc op.cit.
45 Lukáš op.cit.
46 Janáč and Ira 2018.
47 Jakl 2001.
48 Lukáš op.cit.

of the town and the self-assertion of the local community. This cultural resilience drew from a strong and stabilized image of the local heritage, and was manifested in various forms. The resource of the revival (and survival) of Vysoké as a town based on tourist economy had to a great extent to do with the mobilization of heritage. While it would be difficult to prove such assertion in quantitative way, the following cluster of four loosely connected examples provides some circumstantial evidence to this effect.

Built environment heritage

The strong tradition in Czech society of owning a second house created a demand for new houses and apartments in areas attractive for recreation in the Krkonoše[49]. After 1989, neoliberal government policies together with land restitutions and privatization weakened regulatory mechanisms, leaving the space relatively free for developers. Up to 2011, Vysoké remained the only major winter sport centre in Krkonoše Mountains without newly built apartment houses. Other ski resorts in vicinity of Vysoké, such as Rokytnice, Harrachov, Mísečky or Paseky, have changed significantly under the pressure from developers and experienced the mountain alternative of urban sprawl. Vysoké for a long time resisted this trend, especially due to the disapproving attitude of the local authorities and the community, which tend to see such proposals as a threat both to the community and to the town heritage rather than a business opportunity.

Only in 2008 did a new development project start close to the ski slope. After prolonged negotiations, the municipal authorities accepted the plan to build a resort resembling a historical village, based on the "traditional" architecture and modeled after the Vysoké square before the fire of 1834. While references to the idealized concept of the "Czech hut" as a powerful constituent of Czech national myth[50] and motifs of vernacular architecture have been present in Czech countryside architecture since the late nineteenth century, such a large scale imitation of historical settlement remains unique in the regional and probably also

49 Vágner and Fialová 2004.

50 Woitsch 2008.

the national context[51]. The investor in the project even consulted representatives of local heritage institutions about the design, especially the director of Vysoké Museum[52] (Fig. 5).

Thus, the power of local identity and heritage significantly shaped the project, which went beyond a simple business-oriented heritagization. The entire resort has been named after the statute of Good Shepherd which stood in Vysoké square until the fire of 1834. Now, its copy stands at the entrance of the new resort.

Fig. 5. Recent development project "At the Good Shepherd" in Vysoké nad Jizerou. Source: Website of the developer – http://www.sachtyski-realitni.cz/galerie.

[51] Škabrada 2015.
[52] Měchura 2015.

From the perspective of national heritage authorities and experts, such heritagization represents a threat to authenticity. Experts complained that similar "caricatures" of vernacular architecture have appeared in last two decades especially in regions with some form of legal regulation (heritage areas) protecting their "character" and "landscape"[53]. Local stakeholders in Vysoké to a certain extent shared such perspective and dubbed the resort in a derogatory way as "the Praguers' town," partly in order to emphasize its fake character. However, in the end they decided to accept the new development. The problem of "authenticity", defined in art-history terms, apparently did not pose a major problem for the local community. Declared to be inspired by the images of Vysoké before the fire of 1834 and other traits and references to local community heritage, this made the resort in a way a materialization of the local heritage.

Performing heritage

The tradition of amateur theatre in Vysoké, spread over more than two centuries, is generally recognized by the local community as a testimony to its urban and cultural character and an elementary constituent of local heritage. However, it received the official status of intangible heritage and associated entitlement to be protected and supported by the state only in 2012 (puppetry) and 2017 (amateur theatre)[54]. The Krakonoš Theatre (named after the mythical guardian/spirit of the Krkonoše Mountains) and its building represent a crucial element in community life of the town and the festival of rural amateur theatres hosted here annually since 1971 is perhaps the most visible feature and defining symbol of Vysoké from the national perspective.

The survival of the theater and especially of the festival in the 1990s represent an illustrative example of cultural resilience. The disintegration of the institutional supporting network, ranging from the disbandment

53 Škabrada op.cit.: 244.

54 Ministry of Culture of the Czech Republic 2019.

of amateur theatre groups associated with agricultural collective farms in rural areas, general loss of interest in amateur theatre following the 1989, and transformation of regional and national cultural policy and institutions, made the organization of the event already in the autumn of 1990 almost impossible. However, mobilization of local resources and volunteer work of people from Vysoké (including financial sponsorship) secured the survival of the festival. As the contemporary director (and granddaughter of a former director from the interwar period) puts it: "it was a question of honour to keep the festival alive and preserve the legacy of ancestors". Throughout the difficult period of the 1990s the festival relied on steady support from the municipal authorities and large turnouts, by which the local community (allegedly always around 50 % of the audience) demonstrated its engagement in such "heritage" activity.

The winter sports heritage, most notably the Nordic skiing, remains omnipresent in Vysoké. One of the four rooms of the museum is devoted entirely to the local history of skiing, and the motto "town of the most beautiful winter" is the headline of the town's website. Historical photographs of skiers and ski-jump hills decorate the walls of the restaurant on the main square. Despite the fact that the local ski club is still active, and cross-country skiing trails (depending on the natural snow level) and slopes for alpine skiing (with artificial snow) are available in the town, Vysoké presents itself as a cradle of skiing in Czech Republic rather than a contemporary ski resort.

This is partly because of the fact that the rich winter sport history overshadows its current state. Formerly a national centre of skiing, from the 1970s Vysoké's role significantly diminished. Ski jump hills with wooden towers which hosted various national championships between the 1930s and the 1960s were torn down in the early 1970s and have never been replaced. Vysoké hills became by then too small for proper ski jumping competitions as well as alpine skiing, and they were located too low to guarantee enough snow for cross-country events. Local competitors, leaders of national rankings until the 1960s, vacated their positions, and local ski manufacturers (Bartoň) could not compete with industrial producers.

Thus, the only major skiing event taking place in Vysoké nowadays is a heritage race. Every other year a cross-country ski race in historical costume takes place at the cross-country skiing resort Šachty. The short distance race, known as "old-style" (*závody po staru*) is allegedly (there is no available historical evidence) a tradition established in the years of the Second World War, when local skiers where not allowed to organize annual races. Instead of a sporting event, they organized a social gathering in historical costumes to disguise the fact that it was a race. After the war, the first old style race took place in 1963. During the early 2000s it became a regular part of the Vysoké winter and slowly turned into a major tourist attraction, regularly attended by more than 1000 viewers and almost 200 competitors, mostly coming from the region. The race has been continuously organized by the local ski-club.

The Club of Czech Tourists in Vysoké also organizes "heritage" events: an annual commemorative hike in the country of forgotten patriots ("*Pochod krajem zapadlých vlastenců*", since 1975), and a ski race "*Ve stopě první středoevropské padesátky*" (In the tracks of the first Central European Fifty-Kilometre), a cross country skiing event commemorating the first international 50 km long race held in Vysoké in 1905 (for the first time held in the 1970s).

Proper "Heritagization"

Since 2015 a local NGO aiming at the development of local tradition "Městečko na dlani" (Little Town in the Palm of Your Hand) took over the organization of the annual funfair and changed it into a community event in which the vast majority of local clubs and inhabitants participate (including the museum). The funfair, formerly a sort of a traveling amusement park, now offers a wide range of products of local folk art and handicrafts, and local cuisine. As such, it captures a snapshot of local identity and serves as a representation of the contemporary image of Vysoké and its past (and heritage). The festival is opened by the mythical guardian of Krkonoše Mountains, Krakonoš, and people wear historical costumes from the late nineteenth and early twentieth century.

This could be a sign of the belated arrival of heritagization as commodification of "rural authenticity" as a strategy for development of rural areas[55]. The NGO was established in a response to the lack of community building activities on the part of the existing heritage institutions,

55 Hájek op.cit.: 560.

notably the museum. The museum experienced its heyday in the late 1980s, when plans for expansion were laid and the institution enjoyed generous support from regional and national authorities. At that time, the average annual turnout exceeded the capacity of the institution (according to employees of the museum at that time). However, from 1990 it dropped significantly, and the museum had to fight for survival. Due to the lack of funds, it transformed itself into a conservation institution which failed to communicate actively with the local community. Occasional lectures and exhibitions did not attract much attention and the local heritage remained largely enclosed behind the walls of the museum. The main installations remained the same as in 1980s focusing on skiing and the culture of the nineteenth century Highlanders. The NGO aims to stimulate the heritage awareness of the community by entering the public space and developing various forms of participative activities reflecting the local heritage. The funfair represents their major achievement so far, but the reconstruction of the historical gate to the park is also a rather telling example of perhaps implicit effort at reinforcement of Vysoké's historical urbanity.

Natural heritage?

When the Czechoslovak government after long preparation finally declared Krkonoše Mountains a nature conservation area in 1963, Vysoké remained outside the territory of the Krkonoše Mountains National Park, rather naturally from the contemporary point of view, which saw the cultural landscape as the opposite to wilderness conservation. However, since 1980 and the rise of global environmentalism, discussions over the sustainability of human-nature relations brought the interests of nature-conservation and cultural heritage closer together. The concept of "cultural landscape" entered heritage discourse in the late 1980s, with the aim of protecting the "mixed" (natural and cultural) heritage, and providing a platform for sustainable development and management – that is, to protect them from over-touristification and other invasive and destructive intrusions into "landscape character"[56].

Vysoké and the surrounding hilly area, with its rural character and traditional architecture, and a strong local identity, would rather naturally

[56] Jacques 1995.

fit the agenda behind any of the various forms of cultural landscape protection introduced in the Czech Republic since 1987[57]. However, the above-mentioned shocks of the early 1990s made any threat of over-tourism or overdevelopment in any other aspect virtually unimaginable. Only in the early 2000s Vysoké together with the regional authorities of the Liberecký kraj developed a project for the protection of the landscape character of the area, which, despite significant local support, was not successful[58].

The reason behind the failure of the Vysoké proposal was, to a certain extent, due to a parallel development of another, similar initiative. By the early 2000s UNESCO introduced the concept of a "geopark", a protected area designed to promote "a holistic concept of protection" integrating "biodiversity and culture" in areas of global geological significance. Vysoké here joined a concerted effort of municipalities and villages promoting the "Bohemian Paradise UNESCO Global Geopark" – an initiative driven by communities from what is known as the Bohemian Paradise (an extended area of sandstone rocks and extinct volcanos), municipalities and NGOs, trying to protect the countryside from disorganized development of tourist infrastructure. The proposal was successful and Vysoké thus entered the world heritage list, if only its geological part.

The southern part of Vysoké nad Jizerou thus now belongs to the area of the UNESCO global geopark Bohemian Paradise/Český ráj (map below, UNESCO park in yellow). Vysoké, however, is relatively distant from the core area of Bohemian Paradise and the final delimitation of the geopark only covers the southernmost corner its area. A decade after the declaration, it is virtually impossible to find any reference to the park on Vysoké's website, or anywhere in the town. The local community never accepted its affiliation to the "Bohemian Paradise" – as it did not fit the local identity.

Until now Vysoké borders with the Protected Landscape Area Jizerské hory from the north, Krkonoše National Park and UNESCO Biospheric Reserve on the east, and the Bohemian Paradise on the south (Fig. 6). The town, however, associates itself only with the Krkonoše – predominantly due to its historically developed local memory.

57 Šantrůčková et al. 2018.

58 Brychtová et al. 2005.

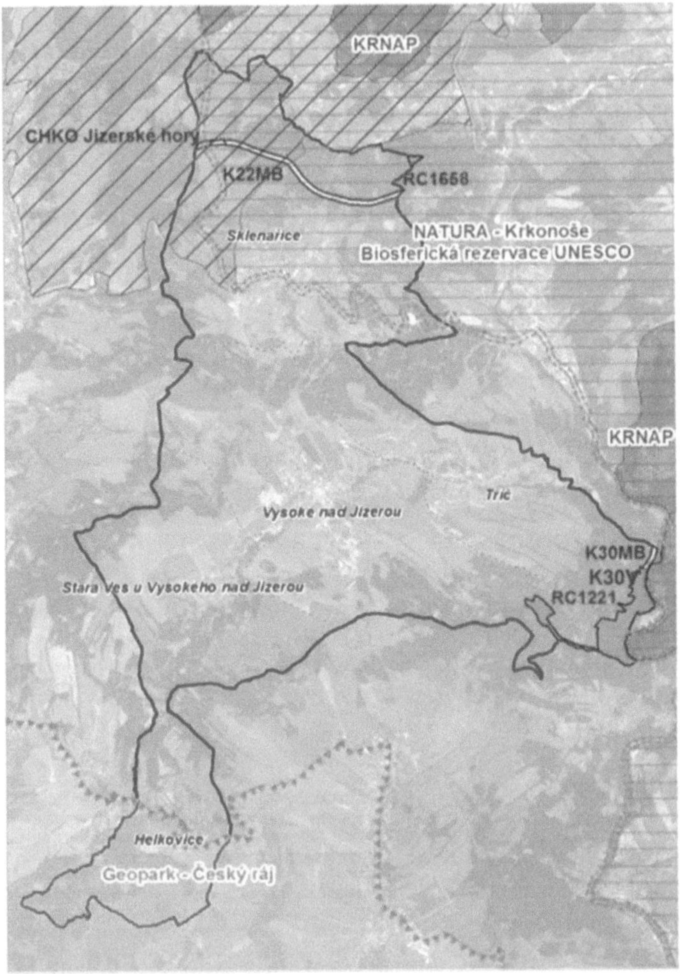

Fig. 6. Territory of Vysoké nad Jizerou (dark line) and protected natural areas as of 2015. Source: Vyhodnocení vlivu územního plánu Vysoké nad Jizerou na životní prostředí (2016).

Conclusions

The story of Vysoké's post-1989 heritagization clearly illustrates the power of local community heritage to sustain the community and its "landscape". Strategies developed to fit the tastes of potential consumers, based virtually only on the economic understanding of heritage and its commodification, and endangering or significantly altering the identity of locals, did not deliver the expected goals. The case of Vysoké makes it clear that especially in small towns local particularities significantly affect the variability of heritage mobilization and management. Tacit rejection of UNESCO natural conservation efforts by local community as well as its protection of the town from urban sprawl development in the 1990s and 2000s clearly illustrate the point.

Similarly, the strong identity of the town and consensual definition of community heritage based on 1) the identity of Highlanders, 2) the urbanity of Vysoké as a cultural centre and 3) the memory of the dawn of winter sports in the Czech Republic, significantly contributed to further re-production of such heritage traits and served as instrumental tools for the survival of Vysoké as urban community after the shocks of the 1990s.

Acknowledgments

This work was supported by the European Regional Development Fund project "Creativity and Adaptability as Conditions of the Success of Europe in an Interrelated World" (reg. no.: CZ.02.1.01/0.0/0.0/16_019/0000734).

Bibliography

Bartoš, M. (2000). Kapitoly z dějin turistiky a lyžování v Krkonoších, stručný přehled krkonošské historie. *Ročenka Státního okresního archivu v Trutnově* (pp. 109–181). Státní okresní archiv Trutnov.

Beel, D., et al. (2017). Cultural Resilience: The Production of Rural Community Heritage, Digital Archives and the Role of Volunteers. *Journal of Rural Studies*, 54, 459–468.

Bollig, M. (2014). Resilience – Analytical Tool, Bridging Concept or Development Goal? Anthropological Perspectives on the Use of a Border Object. *Zeitschrift für Ethnologie*, 139(2), 253–279.

Dědovský, D., Dušek, L., Jandová, F., & Patočka, F. (2019). *Lidová kultura západních Krkonoš ve světle Národopisné výstavy československé 1895*. Správa Krkonošského národního parku ve spolupráci s městem Vysoké nad Jizerou.

Hájek, T. (2002). The Development Potential of Czech Rural Areas and Rural Tourism. *Agricucltural Economics*, 48(12), 559–562.

Halbwachs, M. (2008). *On Collective Memory*. University of Chicago Press.

Holtorf, C. (2018). Embracing Change: How Cultural Resilience is Increased Through Cultural Heritage. *World Archaeology*, 50(4), 639–650.

Ira, J. (2013). *Local and National Identity in Historical Monographs on Bohemian, Moravian and Galician Towns (1860–1900)*. Dissertation, Charles University, Prague.

Jacques, D. (1995). The Rise of Cultural Landscapes. *International Journal of Heritage Studies*, 1(2), 91–101.

Jakl, M. (2001). Mariánský sloup na náměstí Vysokého nad Jizerou ve vybraných souvislostech sochařské výzdoby vysocké krajiny. *Z Českého ráje a Podkrkonoší*, 14, 191–212.

Jakl, M. (2004). Viktor Dyk a Vysocko. In: J. Tomeš (Ed.), *Básník a politik: Sborník z konference k sedmdesátému výročí úmrtí Viktora Dyka* (pp. 118–123). Akropolis.

Janáč, J. & Ira, J. (2018). Materializing Identities in Socialist and Post-Socialist cities. In: J. Ira & J. Janáč (Eds.), *Materializing Identities in Socialist and Post-Socialist Cities* (pp. 11–29). Karolinum.

Kibic, K. &Vošahlík, A. (2012). *Památková ochrana a regenerace historických měst v České republice: 1945–2010*. Národní památkový ústav.

Klapka, P. & Martinát, S. (2005). Geografická analýza vybraných populačních a sídelních charakteristik Krkonoš (1850–2001). *Opera Corcontica*, 42, 139–152.

Klíma, A. (1993). "The Czechs." In: M. Teich & R. Porter (Eds.), *The National Question in Europe in Historical Context* (pp. 228–246). Cambridge university press.

Klusáková, L., et al. (2017). *Small Towns in Europe in the 20th and 21st Centuries. Heritage and Development Strategies*. Karolinum.

Krejčí, M. (2012). Památková péče od normalizace k transformaci. In: I. Lorencová (Ed.), *Věda a technika v Československu od normalizace k transformaci* (pp. 239–250). Národní technické museum.

Lenderová, M., Jiránek, T., & Macková, M. (2013). *Z dějin české každodennosti: život v 19. století*. Karolinum.

Lukáš, V. (1992). *Vysoké nad Jizerou. Pohledy do minulosti dávné i nedávné*. Město Vysoké nad Jizerou.

Macura, V. (2015). *Znamení zrodu a české sny*. Academia.

Meerow, S., Newell, Joshua P., & Stults, M. (2016). Defining Urban Resilience: A Review. *Landscape and Urban Planning*, 147, 38–49.

Němec, L. & Zusková, I. (2005). Změny sněhové pokrývky v České republice od roku 1926, *Meteorologické zprávy*, 58, 135–138.

Průková, M. (2007). Kramářova vila ve Vysokém nad Jizerou. *Acta Musei Nationalis Pragae, Series A – Historia*, 61(3–4), 55–60.

Rais, K. V. (1894). *Zapadlí vlastenci*. F. šmáček.

Šantrůčková, M., Kučera, Z., & Chromý, P. (2018). Peripheral Cultural Landscapes: Their Preservation and Regional Distribution in Czechia. *Životné prostredie*, 52(4), 228–232.

Škabrada, J. (2015). Ke kořenům současné výstavby pseudotradičních budov na českém venkově. In *Odkaz Národopisné výstavy českoslovanské 1895 v oblasti etnografie, muzejnictví a památkové péče* (pp. 233–244). Národní zemědělské muzeum.

Skrbek, J. (ed.) (1936). *Zápisky vysockého Omladináře, politického vězně F. M. Hájka ze Svatováclavské trestnice v Praze*. J.Skrbek.

Steinführer, A., Vaishar, A., & Zapletalová, J. (2016). The Small Town in Rural Areas as an Underresearched Type of Settlement. Editors' Introduction to the Special Issue. *European Countryside*, 8(4), 322–332.

Štemberk, J. (2013). Cestovní ruch a česká společnost na přelomu 19. a 20. Století. *Historická sociologie*, 14(1), 75–88.

Štulc, J. (2012). Debates on Urban Conservation Policies in Communist Czechoslovakia of the 1960s and 1970s: A Note of Warning in the Context of the Present Unesco Discussion on Historic Urban Landscape. In: J. Purchla (Ed.), *The 1st Heritage Forum of Central Europe* (pp. 91–100). Międzynarodowe Centrum Kultury.

Swain, N. (1999). Agricultural Restitution and Co-operative Transformation in the Czech Republic, Hungary and Slovakia. *Europe-Asia Studies*, 51(7), 1199–1219.

Vágner, J. & Fialová, D. (2004). Prostorové rozmístění druhého bydlení v Česku – základní charakteristiky. In: J. Vágner & D. Fialová (Eds.), *Regionální diferenciace druhého bydlení v Česku* (pp. 49–56). Katedra sociální geografie a regionálního rozvoje Přírodovědecké fakulty Univerzity Karlovy v Praze.

Vaishar, A. (2004). Small Towns: An Important Part of the Moravian Settlement System. *Dela*, 21, 309–317.

Vaishar, A., Šťastná, M., & Stonawská, K. (2015). Small Towns – Engines of Rural Development in the South-Moravian Region (Czechia): An Analysis of the Demographic Development. *Acta Universitatis Agriculturae et Silviculturae Mendelianae Brunensis*, 63(4), 1395–1405.

Vaishar A., Zapletalová, J., & Nováková, E. (2016). Between Urban and Rural: Sustainability of Small Towns in the Czech Republic. *European Countryside*, 8(4), 351–372.

Vařeka, J., & Plessingerová, A. (1991). Vzpomínka na vysocké prázdniny univ. prof. dr. Čeňka Zíbrta. *Český lid*, 78(3), 213–215.

Vodseďálek, I., Sládková, A., & Řehořová, I. (2004). *Putování s medvědem. Sborník vydaný k 650. výročí města Vysokého nad Jizerou. 1354–2004*. Nakladatelství Bor.

Vystoupil, J., Kunc, J., Šauer, M., & Tonev, P. (2010). The Development of Tourism and Its Spatial Organization in the Czech Republic Between 1990 and 2009. *Urbanismus a územní rozvoj*, 13(5), 93–108.

Woitsch, J. (2008). Ta naše chaloupka česká? Proměny mýtu lidové architektury. *Dějiny a současnost*, 30(6), 30–33.

Zemanová, J. (2012). *Vývoj struktury obyvatelstva malých měst v České republice v letech 1869–2001*. Diploma thesis. West Bohemian University.

Zhang, X. & Li, H. (2018). Urban Resilience and Urban Sustainability: What We Know and What Do Not Know? *Cities*, 72, 141–148.

Documents

Brychtová, J. et al.(2005). Vysoké nad Jizerou – Roprachtice. Studie k záměru vyhlášení přírodního parku. https://www.kraj-lbc.cz/public/ozivpr/studievysoke_a56f50e830.pdf.

Brynda, E., Dostál, J., Pater, B. (1938). *Padesát let klubu československých turistů*. Klub československých turistů.

Czech Statistical Office. (2015). *Historický lexikon obcí České republiky – 1869 – 2011*. https://www.czso.cz/csu/czso/historicky-lexikon-obci-1869-az-2015.

Měchura, J. (2015). *Návrat Dobrého pastýře do Vysokého nad Jizerou*. Promotion materials of the new resort. http://www.sachtyski-realitni.cz/clanek_awm_dwn.pdf?root=Li4vLi4v&data=jHc3cmG1lekN2LBuVmu4OIyZ-fOZvumJvk/TiXF7IC+gsLzgLcoW1MqXYsl8DNtiYBsw4YDAXeHnT.

Ministry of Culture of the Czech Republic. (2019). Seznam nemateriálních statků tradiční lidové kultury České republiky. https://www.mkcr.cz/seznam-nematerialnich-statku-tradicni-lidove-kultury-ceske-republiky-299.html.

Národnostní mapa Království Českého na základě sčítání lidu z roku. (1880). In *Virtualní mapová sbírka Chartae-Antiquae.cz*. Výzkumný ústav geodetický, topografický a kartografický, v.v.i. http://chartae-antiquae.cz/en/maps/18189.

UNESCO, NICH. (2016). *International Expert Meeting on Cultural Heritage and Disaster Resilient Communities: Within the Framework of the Third Un World Conference on Disaster Risk Reduction (WCDRR): Proceedings, 11–13 March 2015, Tokyo, Japan, 14–17 March 2015, Sendai, Japan*. NICH.

Větrník (2017). Municipal newsletter of Vysoké, published monthly since 1968. https://www.vysokenadjizerou.cz/vetrnik/.

Vyhodnocení vlivů územního plánu Vysoké nad Jizerou na životní prostředí. (2015). https://portal.cenia.cz/eiasea/download/U0VBX0xCSzAzNEZfdnlob2Rub2NlbmlfNTQ1NDQyODcxMjM2ODA3NzQ2OS5wZGY/LBK034F_vyhodnoceni.pdf.

Vysoké nad Jizerou. (2015). *Program rozvoje města Vysoké nad Jizerou na období 2016–2020*. https://www.vysokenadjizerou.cz/assets/File.ashx?id_org=18830&id_dokumenty=1270.

Section IV

Cooperation

Corredor Bio-Comechingones, an innovative regional scale agreement among small towns[1]

Mónica Jimena Ramé

Introduction

In the shadow of the Paris of America, Buenos Aires, or of one of the world's seven natural wonders, the Iguazu Falls, in a peripheral area of Argentina, the Conlara Valley can be found in the province of San Luis. In 2003 six towns from the Valley, each different from the other – Villa de Merlo, Los Molles, Carpinteria, Villa Elena/Cortaderas, Villa Larca and Papagayos – created an innovative territorial network in Argentina: Inhabited but Natural and Protected Area (AHNP) called the Bio-Comechingones Corridor.

The following pages address the emerging and fading of the Bio-Comechingones agreement, whose main goal was to protect the natural, tangible and intangible heritage of the Corridor. At the same time the creation of the network was an attempt to learn how to use it as a resource in a sustainable way, in order to struggle with different regional imbalances identified as: real estate pressure; difficulties in providing basic services; hazards to the environment and the landscape; transculturization and risk of the loss of identity and heritages; rural uprooting; impossibility of adequate individual responses due to the scant structure of each municipality[2].

These territorial imbalances are the result of the economic base of the region being the tourist core of the Comechingones mountains. As a

---

1 This chapter is based on the partial outcomes of the doctoral thesis: "*Las contradicciones de la transculturación de modelos de ocupación urbana a localidades turísticas serranas*".

2 Bio-Comechingones Corredor Agreement 2003.

post-tourism effect the demographic rate is growing due to the amenity migration process. The new inhabitants come from the big cities and metropolises of Argentina and bring with them their city lifestyle, which establishes a tension with the mountain lifestyle. Therefore, it is necessary to develop a new balance between the locals and the migrants and preserve the identity and the aboriginal heritage and even prevent the possibility of a natural disaster – the destruction of the environmental factors which generate the unique microclimate of the area.

Over the past few years a wide variety of heritage strategies of the agreement have been applied, such as the development of strategic regional plans, the recovery of the aborigine and revitalization of the town's historical centre and public spaces. Unfortunately, because of lack of appreciation from central government the agreement lost its financial support and could not continue working. This led to the loss of the Federation of French National Parks management consultancy. However, this unfinished agreement left as good practice among the towns the need to think jointly about the issues that they share.

Being aware of the value of these strategies, it is important to interpret and rescue the undercover theoretical approach that underpins them. To achieve this goal, the methodology to be used includes the harvesting of data with structured surveys, unstructured interviews with key actors and a comprehensive bibliographic review on issues such as the historical backgrounds of the case study region and conceptual frameworks on cooperative networks. Thus, it allows a descriptive model to be built that offers theoretical explanations as to how the Corridor's localities interact, how the flows occur between them, and how it tries to balance the Bio-Comechingones system.

The Bio-Comechingones Corridor

In the Argentinian province of San Luis, 6 towns located on the western slope of the Comechingones Mountain decided in 2003 to crystallize the common history that united them by signing the Bio-Comechingones Corridor agreement. The agreement addresses issues that go beyond historical aspects; its objectives involve political-economic aspects, the impact of urban growth and environmental protection. In this way, a linear territorial distribution network 42 kilometres long was formed, which begins at the northern end of the province of San Luis, a point bordering the province of Córdoba, and ends southeast of the Conlara Valley, towards the locality of Punilla, aligned on the provincial route nº1. In a North-South direction, the localities that make up the network are: Villa de Merlo, Carpintería, Los Molles, Cortaderas-Villa Elena, Villa Larca y Papagayos (Fig. 1).

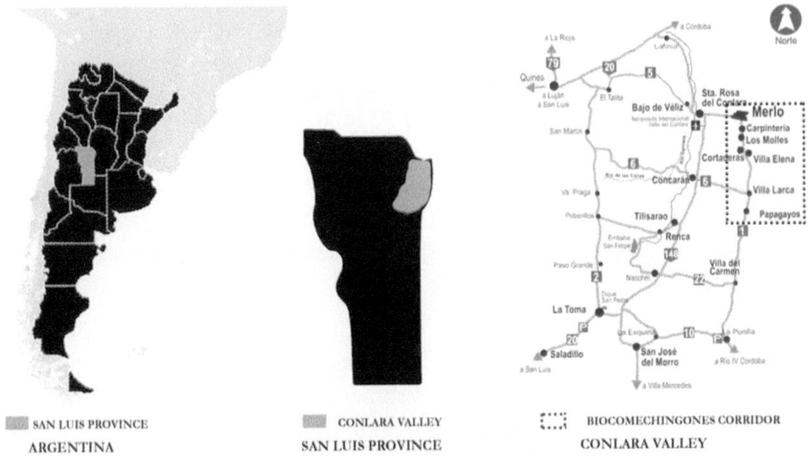

Fig. 1. Map of the Corridor locations according to different approach scales. Source: Own elaboration based on data provided by web images.

The Corridor is the possessor at each of its extremes of a natural heritage of high value for its unique characteristics. On the one hand, the locality of Papagayos has the conditions that make possible the growth of a type of endogenous palm: Palms Caranday. At the opposite extreme, in Villa de Merlo there is a particular microclimate which is in third place on an international scale[3].

On the other hand, the corridor can be characterized as a receptor for amenity migrants as part of the phenomenon of post-tourism[4], due to its role as a mountain tourist destination. Amenity migrants are those people who, having been tourists in one place, return to become permanent or temporary residents[5]. In particular, in the Bio-Comechingones

---

3 Mercau 1990.

4 Bourdeau 2008.

5 Moss 2006.

Corridor the new inhabitants come from large Argentine cities, such as Rosario, Córdoba and, mostly, from Buenos Aires[6] and bring with them the urban lifestyle, which establishes a tension with the mountain lifestyle (Fig. 2).

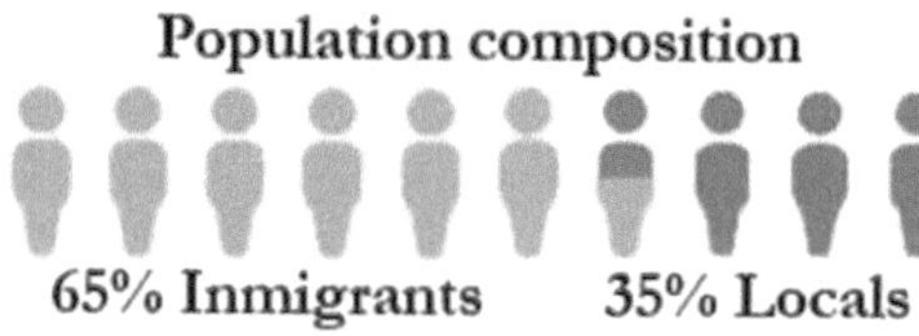

Fig. 2. Percentage of migrants in relation to the total population of the Bio-Comechingones Corridor, 2010. Source: Own elaboration based on data provided by the survey done within the framework of the doctoral thesis.

This in itself implies real estate development in the receiving localities and heavy pressure to convert the land surrounding the urban centres to a different use, thus helping the developers to obtain urban income, and causing territorial imbalances on a regional scale.

The localities are aware of the value of the natural heritage as well as the reality of urban growth as a result of the amenity migration process. For this reason, they proposed the creation of the Bio Comechingones Corridor, which was presented as an innovative figure in Argentina: Inhabited but Natural and Protected Area (AHNP), with the purpose of combating different regional imbalances[7]:

- Real estate pressure;
- Difficulties in providing basic services;
- Hazards to the environment and the landscape;
- Transculturization and risk in the loss of identity and heritages;
- Rural uprooting;
- Impossibility of adequate individual responses due to the scant structure of each municipality

---

6 Information supplied by the survey done within the framework of the doctoral thesis.

7 Bio-Comechingones Corredor agreement op. cit.

The model adopted for the conformation of the corridor was inspired by one implemented in the region of Vercors, of 17,000 hectares, which constitutes the largest metropolitan reserve in France. The project of the French Regional Natural Parks -PNR- has a broad and consensual scope, from the environmental and urban to the economic and sociocultural. Its areas of work include: heritage and culture, energy and mobility, landscape and urbanism, agriculture- tourism and sustainable development.

In the specific case of this paper, it is pertinent to mention the objectives of the planning and landscape area of the Vercors Park's mixed committee. Its main goals are: to balance development projects and environmental, social and economic projects; to promote coherence in the development process and support territorial dynamics. These two objectives are reflected in the Park Charter[8]. Among its strategies are:

- The unity and coherence of the region, which requires knowledge of the park territory, cities and towns, the relationships between different areas of life, throughout history, to adjust the work started with current issues.
- Councils for the planning of operations. The park team can provide advice and support in the design and construction phase of a landscape plan with an architectural and heritage approach. The actors of development, landscape and architecture are complementary during the design phase of the legal structure at the time of work.
- Use of innovative technologies in urban planning.
- Observation of the demographic trend.
- Alternatives to tax models. It seeks the use of co-construction methods and the creation of tools to find answers to land management issues. In terms of landscape and urban planning, the park offers work, for example, in participatory habitat with experts, eco-neighborhoods, in support of municipalities for the construction and completion of their projects in new districts or rehabilitation of neighborhoods existing and the planning of public or sports spaces.

With the support of the French consultants of Vercors Park, the Bio-Comechingones agreement set out as its main objectives the recovery, protection and conservation of the environmental (natural and scenic) and cultural heritage of the region[9].

---

8 Parc naturel régional du Vercors 2015.

9 Bio-Comechingones Corredor agreement op. cit.

More than 10 years after the signing of the agreement, one of the local referents and promoters of the corridor, Rodolfo Raffo, Secretary of Tourism, Development and Investment of the municipality of Merlo during the period 1999 and 2004, explains the current status of the project:

> "The decision to 'freeze the project' had to do with the lack of interest of the provincial government to support a project that was not their responsibility, not formally expressed but clear"
>
> "On the other hand, local governments (...) lacked the political autonomy and management to face the challenge alone. And finally, the French (Parc du Vercors) supported us until they met with the lack of interest / participation of the Provincial Government"
>
> "Regarding the current situation: there is a formal figure: Corridor of the Comechingones that was embodied in the Strategic Federal Plan for Sustainable Tourism by the Ministry of Tourism of the Nation - PFETS MINTUR. This plan, which already has several updated versions, does an excellent job, dividing the area into zones based on matches of supply characteristics, transport links, etc. in corridors, regions, regions to national tourism. The province of San Luis 'adopted' it long after its creation"
>
> "The formal existence of PFETS MINTUR allows the province and municipalities to have the national support of funds and training for posters, small works, etc" [10]

Historical background of the conformation of the corridor

The extension of the corridor has a historical base; its functional and administrative cohesion goes back to the colonial time of the *Mercedes Reales*. Its surface coincides with the area that covered the first non-native settlement that existed in the region. It corresponded to *La Merced* that belonged to Don Bartolomé from 1688, and where his son Tomas Fernández settled, beginning the establishment of the first families of colonizers. This was located on the western slope of the Comechingones, from the Piedra Blanca stream in the north to Papagayos in the south[11] (Fig. 3).

---

10 Raffo 2016.

11 Costamagna and Costamagna 2012.

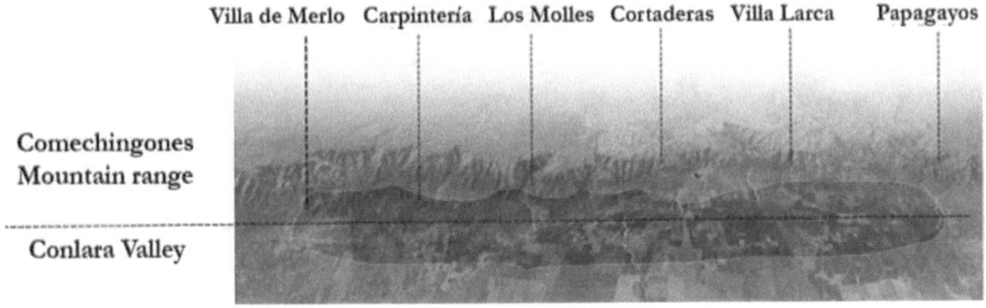

Fig. 3. Extension of the Don Bartolomé *Merced Real*, 1688. Source: Own elaboration based on data provided by Google Earth and bibliographic sources.

This regional tie was maintained beyond the founding of the Villa de Merlo (1797), since each settlement was incorporated as a neighborhood of Merlo, as the municipal radius was extended so that they remained within their common land. This was established, for example, in a letter of 1938 from the Commissioner of the Villa Alberto Menard to the Minister of the Provincial Government, where he requested "1st: to extend the municipal radius (of Merlo) to the place called Los Molles - 2nd: consider the neighbourhood Carpinteria included within the municipal common land. 3rd: authorize the municipality to be able to collect taxes." This was because Merlo provided services to Carpintería, Los Molles and Cortaderas without receiving any remuneration[12].

The taxes that each neighborhood paid, as well as its administrative link with Merlo, ceased when the settlement obtained its legislative autonomy. Such is the case, for example, of Los Molles which obtained it in 1954, Cortaderas/Villa Elena in 1943 and Villa Larca in 1858, although in no case did the strong functional relationship with Merlo end.

Currently, the corridor locations face the challenge of functional cohesion, in an administratively fragmented area, a product of the historical legal organization of the territory. Don Bartolomé *Merced* was divided into two departments: Villa de Merlo, Los Molles and Carpintería belong to the Junín department, while Cortaderas/Villa Elena, Villa Larca and Papagayos belonged to the Chacabuco department. This fact dates from 1885, when the province was divided territorially into 6 different jurisdictional administrative units. In turn, the departments were also

[12] Costamagna 1995.

subdivided into parts following the principle of territorial integrity. The department Chacabuco was divided into five parts, within which the part of Larca covers the towns of Cortaderas, Villa Larca and Papagayos. On the other hand, the department Junín was also divided in 5 parts, being Merlo the head, integrated by Carpintería and Los Molles[13]. This division hinders the fluidity of possible joint projects, which the agreement of the Bio-Comechingones Corridor tries to correct.

The agreement seen as a hierarchical distribution in the cooperative network of localities

The agreement of the Bio-Comechingones Corridor is focused on the principle of agglomeration or synergy. There is a willingness to dismantle a spatial structure concentrated only on the Villa de Merlo, a thriving core of tourist activity in the area, to re-concentrate it in a larger area, which incorporates the six localities. From a conceptual point of view, the aim is to capitalize the formation of an income by positioning at a regional level. The towns adjacent to Villa de Merlo are subject to growing urban pressure, which is verified in the population growth figures and in the appearance of services designed to satisfy and promote tourism (Fig. 4).

13 Costamagna op. cit.

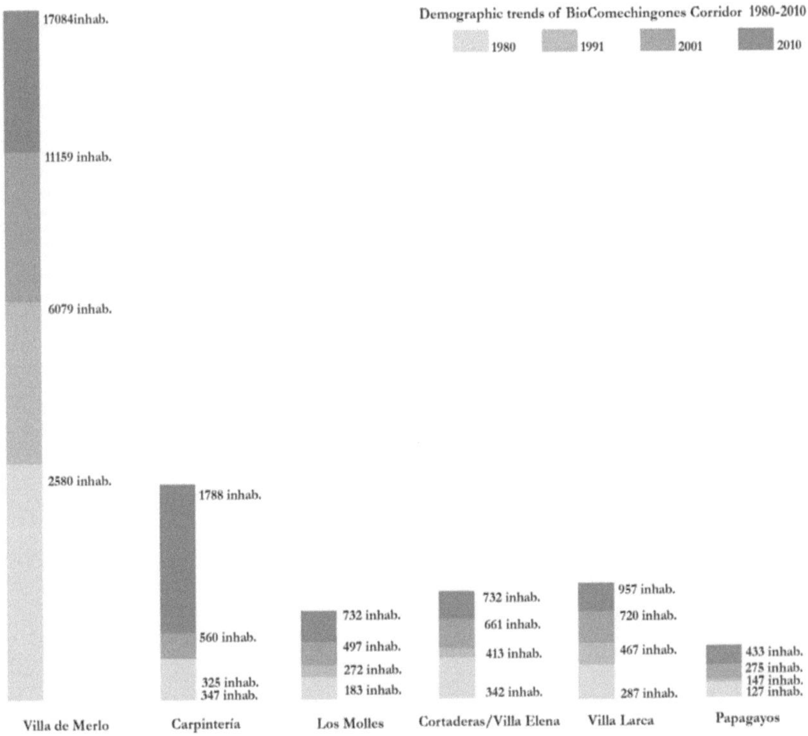

Fig. 4. Demographic trends of BioComechingones Corridor localities. Source: Own elaboration based on data provided by INDEC. INDEC 1980–1991–2001–2010.

In the region, the urban space and the obtaining of income are associated by the hegemonic group with the idea of progress, a meaning that conceals, in a certain way, the negative impact it generates on the natural heritage of the mountain range. Paradoxically, the mountains are the owners of a landscape that is commodified by tourism, either because it is the area where it develops or because they are the landscape to be contemplated. In this sense, the presence of an economy of scale in the tourism sector is assumed in the mountain range slopes that modify both the location scheme and the appearance of agglomeration nodes and sub-poles;[14] economic activity that is accompanied by real estate developments.

From the urban-regional point of view, the agreement seeks to mitigate the progress of targeted urban pressure by reinforcing the construction of the corridor's spatial identity to the exterior as a way of overcoming Merlo's unique spatial identity. To this effect, in this economy of agglomeration of regional scale, each of the localities assumes a defined characteristic according to its offer of services or of environmental/landscape qualities, linked to the shared visual identity that the landscape of the Comechingones mountain range gives them. As a promotion and marketing strategy that sought to diversify the supply of the sector based on the new needs of the market, each one defined an advertising slogan that exalts its most identifiable characteristic:

Villa de Merlo: "3rd World Microclimate"
Carpintería: "Free Flight Capital"
Los Molles: "Heart of Comechingones"
Cortaderas/Villa Elena: "Where Nature beats"
Villa Larca: "The Meeting Place"
Papagayos: "Tradition in Palmeras"

At provincial and municipal levels, the guidance and maintenance of the provision of tourist services as the main economic base has been adopted as a strategic policy[15]. For this reason, the urban space has also been valued through the revitalization of the historic centres of the localities, restoring and normatively protecting the material heritage, and public spaces, both in urban and peripheral areas. The strategic axes of the agreement have therefore focused on the cultural and natural heritage.

Thus, the planning model approaches a third-generation planning model that "rests on two pillars: the cultural enhancement of existing resources and the cooperation of citizens through participatory planning theories (...)are the sum of local values and identities with their benchmarks, which allow product design-this city and far from being the end of the process is the starting point for the testing of a participatory planning model". In contrast to the strategic planning model of the recent

---

14 Camagni 2005.
15 Viecens 2005.

twentieth and early twenty-first centuries, it was a model associated with a phase of intensive capitalist oriented to mass consumption[16].

The slogan that the province used for tourism promotion "San Luis, a happy province" and the comment that the Villa de Merlo was chosen as the ideal tourist destination in the region of the Conlara Valley and symbolic capital can be mentioned as an example of the many initiatives of the state and citizens. Merlo is the locality of *A Place in the World*, the title of an Argentine film shot in 1992 around the town that seeks to rescue the unique landscape characteristics of the mountains and their lifestyle, and of a fragment of the ode to Merlo "natural paradise, divine garden, guarded by the angels of the hill" written by Coco Altamirano, a native of Merlo.

Thus, an idea of the mountain region was created as a picturesque landscape, with colorful and clear skies, based on the documented historical background of its benefits. As early as in 1794, with the aim of founding a new town, the Spanish colonizer Juan de Videla in a first recognition of the area writes "The first, the land of many views and very happy. The second, a lot of water and superior. The third is sixty-four neighbours who have their farms there (…) and lately the beautiful temperament"[17]. Judge Pedáno Alberto Aguilar, on October 28 of the same year, confirms what was said by his predecessor in the Foundation Act, concluding that the most pertinent area to populate a Villa would be Merlo, because "Although all the places mentioned have a view, none equals this place in its good view and plants"[18].

All these associations have a high spatial visual component and configure collective worldviews. The subjective elements, such as the evaluation of the scenic resources or the elements of recreation, merit quantification from the perception. In this sense, John Urry points out that the visual plays a central role in the tourist experience and in "the expectations of what is anticipated will be seen"[19].

With respect to this, from the historical-economic point of view it can be said that these localities are alien to the third phase of the mercantile

---

16 Precedo Ledo, Orosa and Míguez 2010: 13.

17 Mercau op cit.: 8.

18 Mercau op cit.: 9.

19 Urry 2007: 24.

process of the capitalist system, which historically has been given by strata. The first phase occurred during the colonial period -1492/1810-, where the most profitable merchandise consisted of all those minerals that could be extracted from under the earth, mainly gold and silver. The second phase occurred during the nineteenth century or republican-agro-export period, when what was produced by the earth's crust – raw material – formed the main assets exploited. Currently, the third phase of the capitalist system is being experienced, where what is on the earth's surface, the visual-spatial landscape, becomes privileged exchange goods[20]. Or you could also resort to the approach of H. Lefebvre, who mentions that in history a dialectical succession of three great eras can be seen: agricultural, industrial and urban[21].

As has been affirmed, the agreement crystallized the existing urban network and once again unified a territory divided after 400 years of history. Therefore, the corridor is constituted as a network system of a cooperative type, where each of the localities operate articulately as nodes of this and as a whole are owners of a space of high patrimonial value. Descriptive models are intended to offer theoretical explanations of the growth and dispositions of these localities in space, in their different scales -regional, national, world-. From this premise, the concepts proposed in different theories about the network of cities are taken, mainly, those of the Spanish geographer Andrés Precedo Ledo, about the behavior referred to those cities that do not reach the status of global[22].

It is believed, a priori, that in order to address the analysis of this network properly – which continues to be inscribed in a wider network system – it is necessary to divide it into two levels of analysis:

> On the one hand, the inter-network, vertically, the network as part of a global system of cities. For this scale of analysis, it is necessary to apply an approach specific to the current context of urban networks, understanding, in relation to the rank of a city, that currently "the local potential of insertion in international networks appears as a more determining factor than the demographic size or the number of functions"[23].
>
> On the other hand, the intra-network, horizontally, proposes an approach for the study of the internal organization of the traditional and regular network. The concept of rank is observed to place cities hierarchically in

---

[20] Zizumbo and Gutierrez 2015.

[21] Lefevbre [1974] 2013.

[22] Precedo Ledo 2004.

[23] Precedo Ledo op. cit.

relation to the populations they hold and the function they attain, since functionalism is the determining cause of their territorial scale. In this way, the degree of equilibrium that prevails between them is determined. It is also interesting to know the urban primacy index.

Of the multiple definitions that exist on the network of cities concept, Batten has proposed the one that best fits this case study: two or more previously independent cities, and potentially complementary in functions, which strive to cooperate and achieve economies of scale[24]. In this sense, the Bio-Comechingones corridor agreement is an innovative national legal and economic framework for a cooperative network system, especially if one considers that it was formed in 2003. Considering the statements mentioned above, another word emerges, that is key cooperation, which differentiates it from the classic systems where cities competed; cooperation that has its raison d'être in the sense of belonging to the mountain region where they have settled and the shared identity that is awakened.

Networks are feasible, too, to be classified. Precedo Ledo takes four typologies: three serve to characterize this network – the fourth refers to cities that functionally respond to a manufacturing model – based on[25]:

The nature of network externality (Camagni y Salone).

Networks of innovation: they occur between nodes that share a productive orientation and cooperate with each other in a programmed manner (Bio-Comechingones Agreement). They are a type of synergy network. They are composed of centres of high rank, in this case the town of Villa de Merlo and of lesser order, the towns of Carpintería, Los Molles, Cortaderas, Villa Larca and Papagayos.

Type of articulation of urban structures (Dematteis):

Multipolar networks (polycentric) or stable local specialization. The relationship between the nodes is based on relations of complementarity or synergy, and symmetry is not required; there can even be relations of dominance and dependence.

To the radius of the scope of the network. It is the amplitude of the area of influence of the network in the economic and functional space to which it is inserted into the network.

Networks on regional and national levels

---

24 Batten 2004 quoted by Precedo Ledo op. cit.

25 Precedo Ledo op. cit.

The flows that occur between different settlements must be established to understand the relationships between the different population centres of the system to be studied and the degree of integration and cooperation between the centres. These will form a network of nodes with interrelationships that would need more detailed and in-depth study. Likewise, the central place will be established according to the offer of goods and services, as well as their scope.

However, the study will not be reduced to a quantitative view, but rather a qualitative definition of the structuring of the territory. That is, to identify spaces of centrality and hierarchy based on the functions and relevance of the city and how it structures the configuration of the territory.

In turn, the state policies carried out in the area will be identified to create a growth pole, as well as the possible contradictions that the implementation of these measures may entail and the consequences that will result.

A new economy of scale is generated from this described configuration: network economies, which make it possible to face the competitive networks formed mainly by international cities or large national metropolises such as Buenos Aires or Córdoba. The organization as a complementary network, since it is formed by cities of different ranges and complementary functions within the same regional space, allows higher thresholds to be reached than those possible for a single city of this status (Fig. 5).

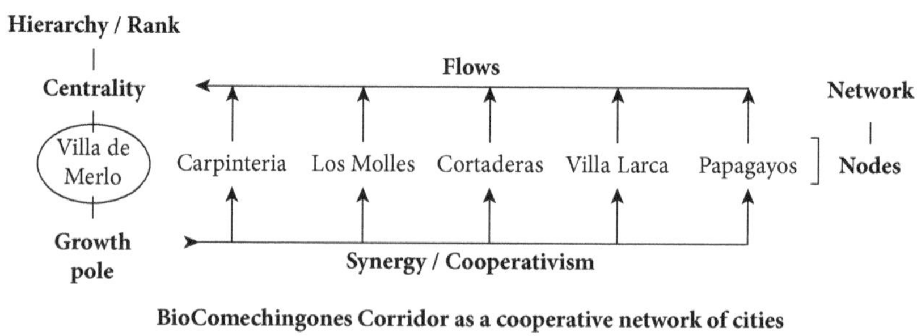

Fig. 5. Conceptual scheme for the operation of the Corridor as a cooperative network. Source: Prepared by the author based on theoretical framework of network of cities.

Conclusion

The Bio-Comechingones Corridor is an innovative figure in the Argentine context that, on the one hand, was able to crystallize an urban history anchored in the colonial period and, on the other, seeks to deal with the territorial imbalances that were identified because of the post-tourism process: urban development and real estate pressure on the mountain range slopes. It is visible mainly in the town with the highest demographic growth of the corridor, Villa de Merlo, where the urban lifestyle of the new inhabitants has been imposed on the local mountain lifestyle: it can be affirmed, then, that the Merlina society is a society of change.

These territorial imbalances find their raison d'être in the economic base of the region with its tourist role initially concentrated by the provincial government only in Villa de Merlo. For this reason, the Bio-Comechingones agreement is a network articulation strategy for regional revitalization and territorial balance, based on a shared identity linked to the mountain landscape and cultural and natural heritage, the result of the search and creativity of small towns. That seeks to protect those towns that have not yet had an exponential demographic growth, of the progressive loss of the link with their past, traditions and cultural identity and landscape- environmental values and at the same time allow urban growth.

However, the vocation to act as a regional unit and a shared identity are not factors that have ensured the desired success of the Corridor. This development strategy that runs between strategic planning and urban marketing, between various jurisdictions and at various state levels, was limited by a provincial political agenda that does not support it with financial or institutional support. Nevertheless, this unfinished agreement was left as a good practice among the localities: the need to think jointly about the issues that go through them, despite their limited capacities.

The Bio-Comechingones Corridor agreement is just another example of those already existing in Argentina of models transcultured from Europe and simply adapted to similar regional realities, perhaps in landscape issues, but completely different in socio-political and economic terms. History must teach that copying must be overcome as a method of developing development strategies and promoting local interdisciplinary actions that aspire as a public policy to the protection of tangible-intangible cultural-natural heritage.

Bibliography

Bio Comechingones Corredor Agreement. (2003). *Convenio Corredor BioComechingones*. Merlo.

Bourdeau, P. (2008). Amenity migration and post-tourism. In: Moss, Glorioso & Krause (Eds.), *Understanding and Man- aging Amenity-Led Migration in Mountain Regions* (pp. 25–32). Banff Centre.

Camagni, R. (2005). *Economía Urbana*. Antoni Bosch.

Costamagna, N. L. (1995). *Historia de Merlo*. O.H.L. Impresos Gráficos.

Costamagna, N. L. & Costamagna, J. M. (2012). Los pioneros del faldeo de los Comechingones. In: Nora L. Costamagna, José M. Costamagna, Águeda Quinteros, Migel A. Padilla, Graciela Santamaría, & Andrea Cardona (Eds.), *Tres Estudios Históricos* (1st ed., pp. 17–95). Payné S.A.

INDEC. (1980–1991–2001–2010). Censo Nacional de Personas y Viviendas. National Institute of Statistics and Census.

Lefevbre, H. ([1974] 2013). *La Producción del Espacio*. Capitan Swing.

Mercau, J. A. M. (1990). *3º Microclima del Mundo, pero… ¿Por qué?* Somma.

Moss, L. (2006). *The Amenity Migrants: Seeking and Sustaining Mountains and Their Cultures*. Wallingford Press.

Parc naturel régional du Vercors. On line 4th Novembre, 2015. http://parc-du-vercors.fr/fr_FR/index.php.

Precedo Ledo, A. (2004). *Nuevas realidades territoriales para el siglo XXI*. Síntesis, S.A.

Precedo Ledo, A., Orosa, J. J., & Míguez, A. (2010). Marketing de ciudades y producto ciudad: una propuesta metodológica. *Urban Public Economics Review*, 12, 13–39.

Raffo, R. (13th January 2016). *Estado del Corredor BioComechingones*. Interviewer Monica Jimena Ramé.

Urry, J. (2007). Introducction. In: Zusman Lois y Castro (Ed.), *Viajes y geografías* (1st ed., pp. 9–29). Prometeo.

Viecens, M. (2005). *Crecimiento y Desarrollo de Villa de Merlo*. Colegio de Arquitectos regional San Luis.

Zizumbo, V. L. & Gutierrez, N. (2015). *La mercantilización de los espacios turísticos tradicionales y su degradación socioambiental*. Educo.

Networking of small communities for heritagization

Paola Pellegrini

Introduction

In recent years, some small communities in the Friuli Venezia Giulia Region in northern Italy attempted actions of heritagization through networking in order first to maintain their resources and cultural identity, and secondly to exploit the possibilities of economic development that international tourism opens up. Their initiatives stem from the lack of appreciation and investment in preservation from the regional government, mostly due to the fact they are sparsely populated areas, at the risk of further de-population as a result of static economy.

This chapter presents two different case studies of heritagization through networking in Friuli Venezia Giulia. They are analysed in their strategies for a new territorial identity, tools for becoming attractive, and the support for local development and impact regarding the territory and tourism.

The first case concerns the initiatives of a group of 11 rural municipalities located in the southern part of the Region along a resurgent river, the main natural feature of the area together with the ancient infrastructures along the water system. Some years ago the communities established a network in order to achieve the critical mass to promote their shared identity and thus support local development. The group tries to capitalize its hyper-local assets and manage them by marketing policies and spatial projects.

The second case study is a project of heritagization through the international networking of two municipalities sharing the same border between Italy and Slovenia; both municipalities are marginal in marginal regions, and share the same landscape, natural features and history.

Thanks to the funding received from the European Union Programme INTERREG ITA-SLO 2007–2013 they started to promote slow rural tourism jointly, and typical products such as red wine and traditional food, thanks to improved marketing of the cultural and natural heritage.

Both cases are under-valued areas which are trying to produce revenues for local communities from "minor" attractions only recently recognized as attractions and resulting from the sum of those of the networking municipalities. Neither case implements a traditional planning approach, either economic or spatial, but a hyper-local approach with the help of experts and local actors. Both processes show how any territorial marketing project must start from the identification of the identity of a territory, in terms of symbols, resources and practices.

A small peripheral region in a strongly characterized country

Friuli Venezia Giulia is a small peripheral region in Italy, 1,217,872 inhabitants in 7,862.3 Km^2 [1]; historically rather poor, mostly relying on non-industrialized agriculture production.

In the north-eastern corner of the peninsula and adjacent to the border of the former Yugoslavia, the region has since the 1990s remained marginal from the impact of the Cold War. In fact the region would have been sacrificed as a battlefield in the event of attack and invasion. The idea of the Friuli Venezia Giulia Region as a place to visit for recreational purposes – in particular the western county of Friuli – was heavily penalized by this geopolitical condition[2].

Not only was the area marginal, but was badly damaged by the earthquake in 1976 which caused almost 1,000 deaths; the earthquake proved to be both a disaster and an opportunity in the 1970s and 1980s, mostly for the large national investments for rebuilding houses and infrastructures. The economic wealth of the region grew substantially and the region contributed to the exceptional flourishing of micro-small industrial enterprises in the Italian north east[3].

---

1 Source: Census Data from Regione Friuli Venezia Giulia, 2018.

2 Basso 2010; Istituto 1978—1986.

3 Giacomello 1988.

Being marginal and strongly conditioned by unfavorable events, but enjoying new-found wealth, the region has since the 1980s struggled to conceive and implement policies for promoting tourism. The appeal of the region is limited because Friuli has mostly "minor" attractions compared with the rest of Italy, meaning that it is not recognized as exceptional and outstanding like Florence or some of the national parks. The challenge is to get it known to both the national and the international public. Friuli is in such a condition that a process of heritagization of resources –cultural or natural, artistic, historic… – is needed in order to exploit the economic advantages international tourism opens up, but also to maintain local cultural identity, which is rather fragile compared to the world-wide known mainstream Italian identities[4].

The reaction following the disastrous earthquake was the first recent process of awareness and heritagization of the local identity, forced by survival needs and shared by the whole regional community; only in recent years some actions for heritagization for tourism enhancement have started to be implemented.

Heritagization of minor assets for tourism development

Italy suffered severely in the economic crisis that started in 2008 and the impact of this crisis does not seem to be contingent, but structural[5]. In times of economic crisis the possibilities that tourism can open up are extremely appealing for local communities with little resources. Tourism, in its multiple forms, has been recognized as an engine for potential economic development for decades[6], but this potentiality remains mostly a statement of principle in Friuli, where the task of promotion is rather difficult because of the "minor" attractions, undervalued first and foremost by the local population.

In fact, the resources to enhance and exploit have to be recognized by stakeholders if they do not benefit from a tradition of being considered heritage, which is very often the case in Friuli. That is to say tourism can become a tool of local promotion only if heritage is considered an asset for re-capitalization by stakeholders. Heritage-asset is not given,

---

4 Marangon, Troiano 2013.

5 Amatori 2017 and Gallino 2015.

6 Antonakakis et al. 2015, Milne and Ateljevic 2001, and Calafati 2000.

in fact, once and for all, but is a dynamic concept[7]: local communities interpret it according to their own contemporary cultural models and use it for their development project. The patrimony thus is a complex social construction, collective attribution of value to objects, buildings, places, streets, cities, and construction of these same artefacts progressively, incrementally in close correlation with life of people[8].

This way, the heritage-asset has no meaning in itself, but when it becomes a resource to ensure the future of the local community – the continuation of the process of social reproduction, that is, the development of productive activities and social practices – it can be transformed[9].

The tourism that appreciates "minor" attractions – such as slow, niche or thematic tourism – is strongly related to this interpretation and has only recently developed into a phenomenon perceivable by marginal areas and able to produce revenues for "minor" local communities. This process of interpretation and heritagization has to be activated and this is what the case studies considered in this chapter tried.

Along the river. The invention of a shared identity

The local networking process

A group of 11 small rural municipalities – average population 3,103[10], average income 11,520 €[11] – decided in 2015 to work together in order to achieve a critical mass to promote actions for the local economic development, basing the strategy on agriculture and tourism (Tab. 1).

The municipalities have in common a river and its water catchment area, the "Stella" from its spring to the Marano and Grado lagoon (Figs. 1 and 2); the river Stella is the Anaxum waterway quoted in Roman times by Pliny, flowing south into the Adriatic Sea, which has maintained the

7 Tosco 2014, Corboz 2001, Babelon, Chastel 1994: 41–53.

8 Pellegrini 2012, Pedretti 1997.

9 Zenghelis et al. 2014.

10 The municipalities in Friuli are very small, the dimensions of the 11 cases considered are average. The municipalities reached 12 with Ronchis, when it joined the group.

11 The municipalities are experiencing either a stable population or a decrease in population, but for the Rivignano Teor case. Udine County, where these municipalities are, has an average income of 13,850 € and a population variation of +3.4 %.

original meandering riverbed in the southern two-thirds of its course[12], a rare occurrence in the heavily transformed North Italian landscape. The extremely differentiated hydrographic system (with lagoon, rivers, streams, bogs, *olle*, ponds, but also artificial canals and ditches) has created woods, riparian vegetation, shrubs and hedgerows, wet and permanent meadows, enclosing an invaluable biodiversity of aquatic and terrestrial flora and fauna. In this land of waters many protected natural areas are recognized[13], because water has historically been a strong bond and its presence prevented sprawling settlements.

These resources are quite unknown to the wider public, but the municipalities share a strong belief in the outstanding value of these natural features and when one municipality decided to promote cooperation and assumed a leading role in the process, Rivignano Teor[14], every stakeholder agreed in the need to conceive actions of promotion.

It is not easy to obtain the agreement of 11 municipalities lead by mayors with different political orientations and the will to cooperate is quite exceptional in Friuli, where more than one attempt by the regional government to encourage unions of very small municipalities had failed in past decades. The main promoter of the initiative, however, the Councillor for Urban Planning in Rivignano Teor, Andrea Pertoldeo, had vision, good skill in networking and a good relationship with the University of Venice IUAV. Without any external input he promoted informal meetings and discussions in 2014, with the support of the Rivignano Teor Mayor and some scholars from IUAV, who took a professional interest in the area and made it the case study of several Design Studios[15].

---

12 The river Stella is a resurgent river that arises at 5 m above sea level, by the confluence of a series of streams born from the water springs, and flows into the lagoon after 45 km. In Roman times the river Stella was an important waterway for the trade of pottery and bricks, allowing fast transport first to the port of Aquileia and then to the lagoon and the Adriatic Sea.

13 Protected Natural Areas are: 7 Habitats, 480 hectares; 3 Areas of Relevant Interest for the Environment – ARIA, along the rivers Stella and Tagliamento; 6 Areas of recruitment, 3,380 hectares; 2 Regional nature reserves, 1,500 hectares; 8 *Natura 2000 Siti di Interesse Comunitario* – ZSC/SIC (plus the Marano and Grado lagoon), 1,690 hectares.

14 Cooperation for administrative purposes between three adjacent municipalities was already active when Rivignano Teor started to promote cooperation.

15 Cafazzo et al. 2013.

The group had to start from scratch, because – apart from Marano, whose territory is officially recognized as attractive because it is mainly occupied by the lagoon of Marano – no specific policies for development and promotion of the area had been conceived, either regional or municipal.

Tab. 1. The municipalities. Source: ISTAT (National Census data, elaboration by the author).

Municipalities	Inhabitants, 1st Jan. '14	Surface (Km^2)	Density (Inhabs./ Km^2)	Population variation 2001–2012	Average income 2011 (€)
Bertiolo	2,538	26.15	97.05	+1.10 %	11,341
Carlino	2,799	30.36	92.19	-0.40 %	12,200
Castions	3,852	32.83	117.34	-3.40 %	11,653
Marano Lagunare	1,900	85.57	22.20	-5.90 %	11,224
Muzzana del Turgnano	2,592	24.32	106.58	-0.50 %	11,569
Palazzolo dello Stella	3,028	34.42	87.96	-1.40 %	11,364
Pocenia	2,578	23.86	108.05	+0.80 %	11,032
Precenicco	1,482	26.98	54.92	-0.80 %	11,727
Rivignano Teor	6,382	47.62	134.03	+9.90 %	11,350
Talmassons	4,127	18.42	224.09	+0.40 %	11,195
Varmo	2,843	42.99	66.12	-1.50 %	12,062
Totale or average	34,121	393.51	100.96	-0.15 %	11,520

The municipalities started the process of conceptual re-capitalization of local assets (not explained in such scholarly terms by any stakeholder in the process) and the potentially recognizable heritage was not confined to what is already evident, such as the natural values of the river Stella and the food products of the local rural tradition, but spans both ancient "castles on the water" and houses belonging to the gentry – according to recent reliable research one of these was the home of the girl whose love story inspired Shakespeare for "Romeo and Juliet" – and archaeological remains, both Roman and mediaeval, both old trees and organic apple production, historical recreational events and water pumps managing waters. The first main goal was to increase awareness in local inhabitants and related entrepreneurship.

Fig. 1. The river Stella and its surroundings in Flambruzzo. Source: The author.

Fig. 2. The river Stella and Villa Savorgnan – Ottelio in Ariis. Source: The author.

The main steps of the cooperation are interesting for the constant inputs given by the promoters and the involvement of two academic institutions; the steps are as follows:

- informal face-to-face contacts among municipalities thanks to the personal networks of the politicians leading the municipalities;
- agreement among 11 municipalities[16] "Stella, Woods and Lagoon. Strategy for a rural territory" (17 June 2015) to define a common strategy of promotion exploiting three elements of the local environment: the river, the woods, the lagoon; this agreement is the first step in the recognition of the main heritage-assets of the shared territory. The goal is to promote an integrated project in order to consolidate real opportunities of development and search for agreements with regional government institutions in order to obtain funding[17]; Arch. Mariagrazia Zatti, Councillor for Urban Planning in Varmo, is in charge of developing the cooperation strategy;
- convention (19 June 2015) and first contract for research (21 December 2015) with the University of Venice IUAV, Department of Building and Conservation (50,000 €, including 30,000 € given by the Region); the municipalities[18] recognized they had neither the skills nor the experience to give content to their proposal, so they asked some scholars to define a "functional and structural organization of the territory apt to promote promotion": the contract is quite vague, which shows how inexperienced the group was in managing the issue[19];

---

16 *Convenzione "Stella, Boschi e Laguna. Strategia per un territorio rurale"* among Bertiolo, Carlino, Castions di Strada, Marano Lagunare, Muzzana del Turgnano, Palazzolo dello Stella, Pocenia, Precenicco, Rivignano Teor, Talmassons and Varmo. The task is defined as "*Un percorso di cooperazione per lo sviluppo di un territorio, all'interno della costruzione di un quadro conoscitivo e interpretativo del territorio come risorsa*".

17 Capulli et al. 2016: 126–158.

18 Not all the municipalities were certain in their support for the convention; one municipality, Ronchis, left.

19 IUAV *Convenzione Istituzionale, La costruzione di un quadro conoscitivo e interpretativo del territorio come risorsa, responsabile scientifico:* P. Grandinetti, *committente*: Comune di Rivignano Teor (Ud), Comune di Palazzolo dello Stella (Ud), Comune di Marano Lagunare (Ud), http://www.iuav.it/Ateneo1/strutture-/architettu/DACC/RICERCA/CONVENZION/CONVENZION/ index.htm, last accessed December 2019.

- recognition of the area governed by the 11 municipalities as three "Landscape Sub-units" by the team drafting the Regional Landscape Plan – PPR[20] and subsequent agreement between the group and the Regional Government to elaborate the studies and the participation process requested by the PPR (25 July 2015); the municipalities showed a clear will to be recognized as a unit and to obtain the funding offered by the regional Government to those which volunteered to take part in the drafting of the PPR (30,000 €). The University of Venice IUAV together with the University of Udine were in charge of this task[21];
- transfer from the Regional Government to the Rivignano Teor municipality of the property of the Renaissance complex - Villa Ottelio - for public purposes (13 November 2015) and the related funding for its restructuring[22]; the municipality proved able to conceive projects and manage its assets, so the Regional Government was more than happy to transfer such large, unused and decaying heritage – source of problems and responsibilities – to the local authority; in 2016 a project for the regeneration of the Villa was drafted by IUAV[23].
- second research contract (28 April 2017) with the University of Venice IUAV, Department of Building and Conservation (15,000 €) for the drafting of the project "*Stella, boschi, laguna. Un territorio naturale e rurale, un'identità come risorsa*" to answer the Call of the EU Rural Development Program 2014–2020 – RDP,

[20] *Accordo con la Regione Friuli Venezia Giulia per lo svolgimento di attività finalizzate all'elaborazione del Piano Paesaggistico Regionale del Friuli Venezia Giulia (quadro conoscitivo e interpretativo del territorio inteso come risorsa, finalizzata alla partecipazione al Piano Paesaggistico Regionale (PPR) [...])*; for this agreement the municipalities were divided into three groups corresponding to three different landscapes.

[21] In synthesis the requirements are: Survey of the Zoning; Survey of the water system; Participation process with facilitators; Description of the main elements of the territory focusing on potentialities and problems. In the path to draft the PPR the group was the main participant of the 11° WORKSHOP, 30 November 2015.

[22] *Cessione a titolo gratuito al comune di Rivignano Teor del compendio immobiliare denominato 'Villa Ottelio' per finalità di pubblico interesse, numero Delibera*: 2251.

[23] IUAV *Convenzione Conto Terzi, Elaborazione di uno studio di fattibilità finalizzato alla redazione di un progetto di completamento e ristrutturazione del complesso di villa Ottelio ad Ariis di Rivignano Teor (UD), responsabile scientifico*: P. Grandinetti, *committente*: Comune di Rivignano Teor, http://www.iuav.it/Ateneo1/strutture-/architettu/DACC/RICERCA/CONVENZION/CONVENZION/index.htm, last accessed December 2019.

Misura 16.7 Strategie di cooperazione per lo sviluppo territoriale. The proposal asks for 3,000,000 €, was first in the first selection phase, won the second phase and is now being implemented.

- the project of an open-air museum along a scenic path in the area of Varmo, highlighting and connecting the country villas of the gentry, the result of a research contract with the University of Venice IUAV[24].
- in the same 3 years 2015–2017, the University of Venice IUAV started 3 research grants on the topic[25].

The first results

This excellent and innovative initiative has produced little implementation. The Universities, indispensable participants of the process given the absence of both economic resources and technical skill in the municipalities, have produced studies not conceived for operational outputs, often generic and with repeated contents[26]. The features of the area interpreted as heritage asset are listed and framed in an excellent introductory argument but not explained in their reasons, their maintained integrity and their potential for development. The idea which inspires the whole reasoning about the area – that is to say, that the marginal condition resulting from the rurality of the places can become a new centrality – is extremely suitable for a place of no development pressure and aligned with consolidated planning approaches[27], but a critical approach able to produce a detailed SWOT analysis seems to be missing. Probably the future funding from winning the RDP will change this impasse and the study for the regeneration of the old centres in the area - made by another team of scholars from IUAV – will help to take this forward.

---

[24] IUAV *Convenzione Istituzionale e Progetto di ricerca con il Comune di Varmo, finalizzato alla conoscenza, alla conservazione e alla valorizzazione delle architetture rurali e del territorio di Varmo, nei suoi elementi di storicità, nell'ambito della ricerca "Architetture e paesaggi rurali", responsabile scientifico:* P. Grandinetti. See also Dal Bò, Zennaro, 2018.

[25] Call http://www.iuav.it/Lavora-con/assegni-in/incarichi-/archivio/2016/3-soggetti/index.htm, last accessed December 2019.

[26] Rivignano Teor asked for more detailed specification, which IUAV added only for the Rivignano Teor territory.

[27] Lancerini 2005 and Pavione et al. 2017.

The relevant role given to land and buildings by IUAV in defining the heritage-asset has to be noted. As the scholars involved are all architects and planners – one of the reasons why the studies are not multidisciplinary as they should have been in order to offer a broader analysis of the existing conditions – the approach echoes what the architect Giancarlo De Carlo used to say: there is nothing that is not generated by the territory, which is the origin, the primary matrix, of all that it encompasses[28]. Therefore every part is correlated with the other parts and cannot fail to establish itself with its whole immediate and profound relations, which are to be discovered continuously and from time to time in every part of the territory one designs.

Beyond the border. Marginal in a shrinking context

The international networking process

Prepotto, Italy, and Kanal Ob Soci, Slovenia, share not only the same border between Italy and Slovenia, but also the same hilly landscape, natural features and history (Figs. 3 and 4). For centuries the area was a place where different cultures could meet and fight, a gateway to Italy, as the invasion by Turks in 1478 and the defeat of Caporetto in 1917 showed. The two communities remained relatively undeveloped and without official contacts for decades; it can be said these places never recovered from the trauma of the First World War fierce battles and the separation during the Cold War.

The Italian municipality is extremely small – 763 inhabitants in 2015, 33.24 Km^2[1] – in a marginal part of the region and in need of actions in order not to shrink further and completely disappear. The Slovenian municipality is 5 times bigger – 5,710 inhabitants in 2015, 146 Km^2[1] – but equally in need of requalification. With woodland, vineyards and rural villages on the hills, the area has to cope with depopulation, economic and social problems due mostly to territorial marginality and lack of enterprises, not dissimilar to other areas in Europe where the agricultural activity is not able to support life necessities in contemporary times.

---

[28] De Carlo 1966.

Fig. 3. View of Prepotto, Italy. Source: The author.

Fig. 4. View of Kanal ob Soci, Slovenia. Source: The author.

The European Union INTERREG ITA-SLO Program was established to support and increase international cooperation between Italy and Slovenia and some experts[29] considered it a good opportunity to obtain funding to promote local production of wine and food. The promotion of the local wine and the related companies together with the whole rural territory is explicitly affirmed in the marketing and communication Plan of the European Project which won funding from INTERREG program, *Progetto* "RURAL" Cross-border tourism development in rural areas and joint promotion of typical quality products of the municipalities Prepotto and Kanal ob Soci[30].

The idea of the project started from a well-conceived and long-term process of heritagization, that is, the initiative related to Schioppettino Wine, an autochthonous grape variety which almost disappeared as a result of Phylloxera disease at the beginning of last century and was re-established with determination in Prepotto in the 1970s. Schioppettino is today a successful product – in 1987 it obtained the denomination of controlled origin DOC in the Schioppettino subzones of Prepotto and Cialla – but the territory which produces it, favored by a particularly favorable microclimate, does not support its marketing, as it has almost no *allure* and recognized attractions. The Italian area is, in fact, part of a larger area famous for its wines, *Colli Orientali del Friuli*, but it is its marginal and neglected part together with the Slovenian side. The Association established to promote the Schioppettino of Prepotto thinks they need to create what the French summarize with the extended concept of "*terroir*", the link between the wine and the territory. Schioppettino, especially thanks to the actions of the Association, is thus a brand that would like to promote and export the entire territory.

With the project, the two municipalities tried to capitalize their hyper-local assets (minor but interpreted as multi-level) and searched for a common identity to be managed by marketing policies and actions involving space; in this effort the idea of tourism remains essential, mostly to attract as many people as possible to a completely forgotten place.

---

29 The main promoters are Dr. Davide Rocchetto and Michele Pavan, President of the *Associazione produttori dello Schioppettino di Prepotto*, the Association of the producers of the Schioppettino Wine.

30 Call n. 03/2011 *Programma per la cooperazione transfrontaliera Italia – Slovenia* 2007–2013, Priority task: 2. Competitiveness and knowledge based society, Project start date: 2012-09-30, Project end date: 2015-03-30, Approved costs: 446,750.32 €.

The more the profitability of the companies and their commercial penetration increase, the more direct repercussions on both Italian and Slovenian territory are increased (e.g. an increase in tourism or economic exploitation of the land). In fact the cooperation between the two realities was not strictly necessary for the achievement of the project goals and so the networking remained weak.

The results

The following results were planned in the project[31]: "a direct effect that can generate new opportunities for small and medium-sized enterprises especially in agriculture and tourism and greater visibility of the territory involved". The main development element the project tried to sustain was tourism, which can be attracted by the quality of typical agro-food production connected with the cultural and natural resources of the combined municipalities. With this goal the project has promoted cross-border events, cross-border marketing with four farmers' markets, one cross-border tourist route between Italy and Slovenia connected to other already established routes (mainly "*La Strada del Vino dei Colli Orientali del Friuli*"; "*La via dei Monti Sacri*"), four info points, each equipped with a multimedia tower with a touch screen, and one large hangar in Prepotto to host public events.

During the events the local communities experienced involvement and enthusiasm, but after the end of the Project RURAL no further initiative for the joint promotion was undertaken, no cross-border event was organized, the web page of the project was never updated, the web page of the Association of the producers of Schioppettino Wine was closed, and the cross-border tourist route vanished. It must be said that Prepotto is too small to promote initiatives without the efficient support of the organization of a European Project and Kanal was not fully involved. It can be said the initiative lacked the critical mass to maintain what started without external support.

Conclusion. The main elements of interest

Both cases are marginal areas, little developed in comparison with the rich north of Italy and Slovenia and not subject to any development

---

[31] Last accessed June 2019.

pressure. Both cases are in a condition of neglect from the regional government, which is understandable in times of economic crisis and reduction of investment for public promotion and infrastructures[32]. Both cases adopt a networking strategy to start processes able to produce revenues for local communities and heritagize "minor" attractions, whose interest has to be claimed and defined[33]. In neither case has a traditional planning approach, economic or spatial, been drafted. Each case starts from the identification of a territory, in terms of symbols, resources and practices and defines a marketing project with the help of heritage experts and stakeholders to become visible in a global context, which comes from the search for a new territorial identity, the result of those of the networking municipalities[34].

The main elements of interest of both processes is the idea of hyper-local and related territorial marketing policies[35]:

- the municipalities realize that a "minor" heritage must be exploited and transformed into a hyper-local asset, meaning that being local makes it extremely original, natural and unspoiled, and unobtainable elsewhere; the goal of this invention process is to become "hyper", that is to say to define a strong identity (hyper-rural, hyper-natural, hyper-specialized); in synthesis super-local assets to create an extra-strong identity[36];
- the newly created network is used to exit the proximity trap and have the world as reference: it is a well-defined community, which wants to produce very specialized information and direct communication about itself not only to the same community, but to the external world in order to penetrate the market;
- the extra-small government bodies recognize that a critical mass is needed to obtain effective results, which is the real and big difference from the past promotion of the territory in Friuli. This trend towards grouping municipalities is expressed by the regional law for reform of local bodies, which obliges small municipalities to get together and

32 Geiselberger 2017.

33 Golinelli 2008.

34 Carta 2006.

35 Pellegrini 2015.

36 Birns 2019.

share services and policies[37]. However, the municipalities studied in this chapter operated on a voluntary basis and anticipated the law.

Some lessons emerge clearly from the processes:

- the level of critical mass needed depends on the context and the quantitative dimension of the territory involved in the action of heritagization can be a minimum but the effort to group stakeholders must be tried;
- in extremely small communities, social capital and personal networking are essential and possible to start a process, and have to be constantly exercised during the process;
- hyper-localism requires a common identity, which can be invented and promoted; constant support with new inputs must be given to the process in order to keep it active until the target stakeholders are capable of self-promotion: the *capability to realize the heritagization process* must be increased[38];
- the processes of cooperation recognize and involve all kind of elements to get the most out of the territory. Their availability comes from both a shared interpretation and a possibility of utilization for the interest of inhabitants. The process therefore has two parts: an assessment must be first made about what can be considered or transformed into heritage, conceptually and materially[39]; secondly, a development project must be imagined critically, considering what local resources can be activated[40];

---

37 The reform (*Legge Regionale* 26/2014) explains the need for unions as a reaction to the economic crisis, to the excessive costs of local bodies operating and to the inadequacy of small towns in providing services. Small municipalities are the absolute majority in Friuli. The law promoter clearly states "Small is no more beautiful" and takes the "Camposampierese" as a virtuous example. The Camposampierese (PD) consists of eleven municipalities, like the group in Friuli, but it is smaller in extent (226 versus 393 kmq), much larger in population (100,000 versus 34,000 inhabitants) and richer in average income (16,000 versus 11,500 €). The comparison helps to understand the low density and lack of wealth in Friuli.

38 Sen 1985.

39 Magnaghi 2000.

40 As an example: if IUAV claims the dismissed and neglected old country complexes along the river Stella are valuable and the development project must use them, the definition of which local resources can be activated is necessary, but these have not been mentioned so far. This is why the initiative and process have been excellent, considering the limited resources of the municipalities, but the material results are disappointing, considering the time-span and the goal of the local governments: "Let's do something real".

- the networking municipalities are facing a challenge: instead of promoting the fact that they are being local as a resource and opening up to the world, they might end up in closing into an extreme localism, proclaiming they are this way defending their prerogatives, as they have usually done for years.

Bibliography

Amatori, F. (2017). *L'approdo mancato. Economia, politica e società in Italia dopo il miracolo economico.* Feltrinelli.

Antonakakis, N., Dragouni, M., & Filisc, G. (2015). How strong is the linkage between tourism and economic growth in Europe? *Economic Modelling,* 44(1), 142–155.

Babelon, J.-P., & Chastel, A. (1994). *La notion de patrimoine.* Liana Levi.

Basso, S. (2010). *Nel confine. Riletture del territorio transfrontaliero italo-sloveno.* EUT.

Birns, N. (2019). *The Hyperlocal in Eighteenth- and Nineteenth-Century Literary Space.* Lexington Books.

Cafazzo, M., Grandinetti, P., & Grandinetti, R. (2013). Ripartire da un fiume: Cronaca e teoria di un percorso di marketing territoriale. *Economia e società regionale,* 2, 213–219.

Calafati, A. G. (2000). Il capitale come paesaggio. *Foedus. Culture, economie e territori,* 1, 26–39.

Capulli, M., Fozzati, L., Grandinetti, P., & Pinosa, P. (2016). Progetto ecostorico Anaxum: Il parco del fiume Stella. In Atti del Convegno Archeologia e parchi archeologici. Esperienze a confronto, Marsilio.

Carta, M. (2006). *L'armatura culturale del territorio: Il patrimonio culturale come matrice di identità e strumento di sviluppo.* Franco Angeli.

Corboz, A. (2001). *Le territoire comme palimpseste et autres essays.* Editions de l'Imprimeur.

Dal Bò, A., & Zennaro, N. (Eds.). (2018). Giornale IUAV, 160.

De Carlo, G. (1966). *Urbino. La storia di una città e il piano della sua evoluzione urbanistica.* Marsilio.

Gallino, L. (2015). *Il denaro, il debito e la doppia crisi spiegati ai nostri nipoti.* Einaudi.

Geiselberger, H. (Ed.). (2017). *The Great Regression*. Polity.

Giacomello, A. (Ed.). (1988). Friuli ricostruzione: 1976 – 1986. Regione Autonoma Friuli – Venezia Giulia, Segreteria generale straordinaria per la ricostruzione del Friuli, Direzione Generale dell'Istruzione, della Formazione professionale, delle Attività e Beni culturali, 2 vol.

Golinelli, C. M. (2008). *La valorizzazione del patrimonio culturale. Verso la definizione di un modello di governance*. Giuffrè.

Istituto per l'Enciclopedia del Friuli Venezia Giulia. (1978–1986). *Enciclopedia monografica del Friuli Venezia Giulia, vol.4, La Storia.*

Lancerini, E. (2005). Territori Lenti: Contributi per una nuova geografia dei paesaggi abitati italiani. *Territorio*, 34, 9–15.

Magnaghi, A. (2000). *Il progetto locale*. Bollati Boringhieri.

Marangon, F., & Troiano, S. (Eds.). (2013). *Sviluppo economico locale e turismo sostenibile in Friuli Venezia Giulia*. FORUM.

Milne, S., & Ateljevic, I. (2001). Tourism, economic development and the global-local nexus: Theory embracing complexity. *Tourism Geographies*, 3(4), 369–393.

Pavione, E., Pezzetti, R., & Gazzola, P. (2017). The role of "Slow Territories" in the development of sustainable tourism. *European Scientific Journal*, 4/ SPECIAL edition.

Pedretti, B. (1997). La democrazia estetica. In B. Pedretti (Ed.), *Il progetto del passato*. Bruno Mondadori.

Pellegrini, P. (2012). La Carta dei valori, metodo e possibilità. Il caso Carnia. *Planum. The Journal of Urbanism – Atti della XV Conferenza Nazionale SIU*, 25(2), 233–242.

Pellegrini, P. (2015). Hyper-local versus traditional. Development strategies in marginal territories. In G. Fini & C. Saibene (Eds.), *Atti della XVIII Conferenza Nazionale SIU, Planum* (pp. 195–203). Planum Publisher.

Sen, A. (1985). *Commodities and Capabilities*. North-Holland.

Tosco, C. (2014). *I beni culturali. Storia, tutela e valorizzazione*. Il Mulino.

Zenghelis, E., et al. (2014). *Uncharted. The New Landscapes of Tourism*. Actar Publishers.

Railway: From global infrastructure to local heritage. The case of the historic railroad Avellino-Rocchetta in Southern Italy

Consuelo Isabel Astrella

Introduction

Ever since their appearance on the world stage, railways have represented an instrument of globalization[1]. The railway infrastructure has always been a means of communication and mutual exchange among different states and local realities, creating the conditions for sharing ideas, knowledge and techniques. The same trains, as well as the tracks and their characteristics, are often the demonstration of this primordial sort of globalization in which sometimes the construction techniques, or the same trains or locomotives, were borrowed or purchased by other nations. If the greatest effects of the advent of the railway on a global scale can be considered an increase in cultural and commercial exchanges, as well as greater ease of movement from one country to another, analysing the influence of the railway on a local scale, some differences could be noticed[2]. First of all, the advent of the railway originally led to a sort of disorientation in small or medium-sized rural realities that had to expand their horizons and get used to the changes that occurred in the territories and the perception of the landscape with a new infrastructure[3]. The advent of the "iron horse" has constituted an element of estrangement in rural contexts that have often remained unchanged for a long time

1 Giuntini 2001: passim. Godoli, Cozzi 2004: passim. Maggi 2012: 11 et seq.

2 Dinhobl 2016: 121.

3 Mainini, Rosa and Sajeva 1981: 102–103.

and in which traditions and ways of living in the territory linked to the agricultural reality of the places had remained unchanged over time[4].

At the same time, however, especially in the Italian context of small local realities, the railway represented a way out of isolation as well as a valid support for the growth of the local economy as shown by the growing demand at the end of the nineteenth century for a direct rail link from the towns and villages in the hinterland[5]. The construction of a series of secondary railway connections, which had their value at the end of the nineteenth century as stopovers to support the more or less large industrial realities of the time, with the modification of economic structures and the concentration of production activities and exchanges in larger realities, has led, in the last decades, to the closure of many of these railway lines that are no longer profitable. The slow process of decommission of these railways has allowed in some cases their recognition as a historical and cultural asset, as evidence of the work of the productive past of the areas through which the railway passed, thanks to the almost unaltered conservation of the tracks and some structures existing along the railways[6]. At the same time the tracks of these railways, crossing places often internally and originally farming areas, have become excellent means for the revaluation and rediscovery of small local realities focusing on their tangible and intangible cultural heritage and on a new perception of landscapes, often natural, offering the possibility of slow railway tourism.

The railway system, in fact, is in itself a heritage with strong testimonial values both historical and cultural, able to make it an asset belonging to the community, thanks to its being strongly rooted in the territory crossed by the railway. These intrinsic characteristics make railways a complex heritage that unites within itself aspects related to the memory of places and the history of technology, tangible and intangible values, landscape and environmental aspects. These aspects are even more visible and recognizable the more the transformations undergone by the railway lines are minimal. In this context an increasingly important role for the preservation of this heritage is played by the undervalued and often abandoned railway sections which move away from the major

4 Battisti 2001: 63.

5 Maggi 1997: 3 et seq. Bernardini 2003: 17 et seq.

6 Maggi 2002: 377–384.

urban centers to connect smaller and more peripheral realities, through less frequented and lesser-known areas, and which go under the name of secondary lines.

Regarding the Italian case, following the reorganization of the management by *Ferrovie dello Stato* which took place in 1990, the closure of many of these so-called secondary lines, deemed to be financially uneconomical, was therefore decided. These closures were avoided in a few cases only thanks to the will and passion of some volunteers and inhabitants of the places developed along the same railway lines.

The reuse of disused railroads for tourism purposes, using, in most cases, restored historical convoys to better appreciate both the territories and the landscapes as the same means of transport and the different type of travel offered by it, does not however constitute a novelty nor an Italian prerogative. Starting from the United Kingdom, from the second half of the twentieth century, almost every European country has recovered sections or entirely disused railway lines, especially in contexts related to small cities rather than to major traffic routes.

However, if in some countries the reopening of disused railways for tourism purposes has been the main aim for the protection of a historical cultural heritage, constituted in this case by the same railway including also the original trains, in other cases the reopening of a disused railway for tourism purposes has become the means for a greater understanding and enhancement of the territories crossed[7]. In the first case, in fact, the territorial repercussions in terms of tourism and the valorisation of the places were secondary and mainly linked to tourism related to the railway world and travel on historical convoys. In this first hypothesis this type of railway tourism appears to be a phenomenon that, although constantly growing in recent decades, remains, as pointed out by several authors, a niche phenomenon more able to attract rail fans and enthusiasts of the railway heritage[8]. In the second

---

7 «*Railways as attraction* [...] *provide a unique world-making lens through which to view heritage and culture.* [...] *Railways can represent a certain past, perhaps described as a time of early adventure, exploration, expansion or colonization.* [...] *Whereas scenery offered from dome cars and open carriages is often a major attraction, the trains themselves are also a magnet for visitors. Unlike the globally mass-produced automobiles and aircraft, vintage locomotives differ from nation to nation in terms of their physical design, as if a reflection of their society and its culture*». Conlin and Bird 2014: 10. Oppido 2014: 232 et seq. Oppido and Ragozino 2015: 32–35.

8 Stefanovic and Koster 2014: 26 et seq.

case, however, the reopening of abandoned or closed sections and the establishment of tourist trains represented a means to rediscover places, hamlets and traditions forgotten or hardly valued. Almost all Italian tourist railways belong to this second category. According to this theory it can be noted that the establishment of a tourist train has almost always in Italy – unlike other foreign contexts more tied to a railway tourism mainly linked to the journey on a restored historical train – a direct link with cultural and folkloric local initiatives, from which the names of the historical trains with tourist purposes often derive (e.g. train of the wine, train of the chestnuts, train of the temples, of the baroque, of the myth etc. …). The tourist railway thus becomes a means to sponsor often hidden and little-known realities, focusing the attention of the traveler on the places crossed by the train, rather than on the historical convoys, which thus becomes a further and collateral incentive to attract tourists and travelers.

In this same process of rediscovering and exploiting the train, a second phase of globalization of the railway system can be noted. A first phase, as previously mentioned, may consist of the same construction of the railway line, which, with the standardization, not only national but often international, of techniques, types of construction, gauges and locomotives, guaranteed greater speed of construction even in more geographically remote and isolated areas[9]. With the disappearance of the purpose, often commercial and industrial, for which such innermost lines had been created, a slow phase of decline until the closure or disposal of that railway, according to the prescriptions of the economic theory, began[10]. The rigid functionalism to which the railway system is subjected imposes many limitations and constraints in the reconversion phase of the infrastructural system in view of its valorisation and of the territory on which it is grafted, especially if it is in the innermost areas. Among the various strategies developed at the global level for the reconversion and upgrading of the railway system, the one that appears to be more competitive for internal areas is the reuse of closed sections as tourist railways, reusing, in most cases, historical convoys and locomotives, often the steam ones. Naturally, this approach, used in Italy later than

---

9 Dinhobl 2016: 121.

10 *«Economic theory tells us that one of the characteristics of large transport infrastructures is functional rigidity, that is, it is only useful for the purpose for which it was created»*. Cuéllar 2016: 37.

in the rest of the world, represents a strategy already developed in other countries. Despite the necessary variations and specificities due to the individual case, it is precisely in this approach that a second globalizing element is highlighted.

The Avellino-Rocchetta S. Antonio railway as a local heritage able to enhance the region

Fig. 1. The Avellino-Rocchetta S. Antonio railway route with its stations (in a larger font) and stops (in a smaller font). Source: Own elaboration.

As is known, there are more than eight thousand kilometres of railways closed to traffic or underutilized in Italy, considering both those owned by the national company, *Ferrovie dello Stato*, and those granted to other companies[11]. Some strategies for the reuse of these disused sections have foreseen their transformation into greenways. In other cases, the preference was to keep the track, reusing it on special occasions thanks to tourist trains aimed at the rediscovery of rural and internal areas. The Avellino-Rocchetta Sant'Antonio railway is included in this last case. In 2017 it was declared one of the eighteen Italian railways for tourism purposes pursuant to Law 128 of 9 August 2017[12]. This provision

---

[11] Rovelli, Senes, Fumagalli 2004: 46.

[12] More information provided at: www.gazzettaufficiale.it/eli/id/2017/08/23/17G00141/sg (July 2021).

acquires a value of exceptionality compared to other foreign realities in which there are no specific national laws for the development of individual railway branches, being the safeguarding of historical railway lines for tourist use a process that starts from the bottom, which sinks its roots in volunteering and associations. At the same time, this provision at national level is the demonstration of the greater attention of some Italian policies and institutions towards the development of more internal contexts, often far from strong tourist pressures and the laws of the market. In addition, this legal recognition also highlights the ability of this railway section to emerge from the often-disadvantageous conditions linked to its geographical location, thanks to the strategies, initiatives and enthusiasm involved by the various local associations.

Unlike other Italian routes for tourist use, closed or abandoned since the 1990s, the Avellino-Rocchetta railway, despite a long period of underutilization that started as early as the end of the 1980s, was closed to traffic only in 2010. Since then, as usually happens in local communities, both on a provincial and regional scale and both in Italy and abroad, a process of involvement and mobilization of the local population was started by volunteers and associations to request the bodies responsible for managing the service to reopen the local transport line. This is, in fact, one of the few railways that crosses the Apennine ridge, serving the innermost areas of three regions: Campania, Basilicata and Puglia (Fig. 1). The process that has seen the mobilization of volunteers and associations in the front line for the reopening of the line is not new to the railway world. The involvement of local actors and the mobilization in a bottom-up process through voluntary groups or associations is a process already successfully tested in the case of the first tourist railways reopened in Britain in the 1960s. However, unlike foreign experiences and some other Italian cases of tourist railways, some factors were disadvantageous for the reopening of this line. Since this railway was owned by *Ferrovie dello Stato*, it was necessary to find a valid economic response that would lead to the reopening of a service which was not in itself very remunerative for the management body.

The Avellino-Rocchetta railway was in fact already in the pre-unification period[13] among the railway projects that served as a link along the transversal between Naples and Foggia, which has always been a crucial route

13 The idea of improving, thanks to railroads, the connections between the various regions of the Bourbon state, in particular of the more inland agricultural and

for the development of trade in the internal provinces of Campania as well as a valid means of transport and exchange of agricultural products from Puglia to Naples. At the same time, however, with its construction, this railway has shown the lack of correspondence between the political choices taken from above and the actual needs of the population. Although the realization of this section was ratified only in 1879[14], distinguished politicians, such as Francesco De Sanctis, were in favor of an improvement of the railway connections in the southern rural areas, and in particular in the province of Avellino, as early as in the 1860s. The railway, which would have partly followed the route of the ancient royal road to Puglia, would not have disrupted the socio-economic balance, but, instead, had been designed to improve and accelerate trade between the internal production areas, mainly agricultural, and the important rail and port junctions, wanting to increase the economy and trade in Southern Italy after the Unification. Unfortunately, the same choice of the railway track, as well as its construction, represented the main criticality, both political and geographical, for the development of the line.

From the political point of view, shortly before its construction in the 1880s there were some proposals for an alternative route to that already approved by the law of 1879 that would have followed the course of the river Ufita rather than that of the river Calore[15]. In this way, the railway track would have cut out a series of districts, reducing significantly the speed of connections between the various inhabited areas and the development prospects of the place. For these reasons the interested populations objected, and the original project was re-established. This would have united Avellino with Ponte Santa Venere (today Rocchetta Sant'Antonio), already an important railway junction to Melfi and Foggia, crossing the valleys of the rivers Sabato, Calore and Ofanto[16]. Because of its considerable length of more than one hundred and eighteen kilometres and the geological characteristics of the soil, in many places of landslide nature, the works planned to overcome the valleys and cross the rivers and mountains

---

production areas, had already been a topic of discussion at the end of the 1850s. Giorgio, Tenore 1998: 155 et seq.

14 The construction of the railway was part of the routes whose realization was ratified by the law of July 29, 1879. Lenzi et alii 1884: 1 e 19. Di Marzo 1885: 5 et seq.

15 Melisurgo 1884: 9 et seq.

16 Some historians believe that on the decision to change or maintain the original layout weighed more the interests of the landowners and the lords of the municipalities crossed or excluded from the railway route.

were made with techniques and materials used in that period only for the most important railway lines and not for the local ones[17]. Also in this sense, the ability of this railway to overcome local borders can be found.

Despite the opening of the line, which took place for its entire length only in 1895, seven years after the beginning of the works, some events contributed to its failure as an infrastructure to support the economy of the internal areas. Among these, first of all the lack of a direct rail connection of Avellino with Naples, thus interrupting what could have been an axis of direct connection between the two seas, and diverting the commercial traffic arriving in Avellino towards Benevento or Salerno. Secondly, the lack of interest of the local administrations joined by the railway that did not equip the stations, which were and are often located in the valleys a few kilometres away from inhabited centers because of the elevation of the ground and the design of the track, of opportune roads and intermodal connections, slowing the transport of goods and people and making the advent of the railway almost vain[18].

The critical issues outlined above became more evident when the hypothesis of the closure of the line was dreaded on the occasion of the centenary of its opening. In fact, for some time the connections were sporadic and the number of passengers was low, as was the quantity of goods transported. Since then, various hypotheses and feasibility studies have been suggested to have the right economic feedback but above all to save one of the few internal railways of Southern Italy. Once again the most accredited hypothesis was that of a railway to support the industry and the production sector, capable of raising and supporting the southern economy. The idea was in fact that of a railway that was more devoted to the exchange of goods at the service of local industries, mainly food and automobile (some Fiat industries were in Melfi and Pratola Serra), (Fig. 2) which had meanwhile developed in the area.

---

17 As asserted by Crugnola, engineer of the *Società Italiana per le strade ferrate del Mediterraneo*, a body prior to the unification of the various railways in the *Ferrovie dello Stato*: «*The Avellino-Ponte Santa Vergine line is notable for several reasons. The technical conditions of the route are those of a railway of local interest but the methods of construction are of a line of primary importance*». Ogliari 1975: 891.

18 Ogliari 1975: 895.

Fig. 2. A train carrying some Fiat cars from the Fiat industry in Melfi passing through the Rocchetta S. Antonio station. Source: The author, August 2017.

These proposals contributed to support the recovery of an area already heavily damaged by the earthquake of 1980[19]. In parallel with this hypothesis, supported by local and national authorities, other local bodies, such as the *proloco*, also proposed a reconversion of the whole route for tourist purposes, focusing more on the beauty of the territory and the landscapes visible from the railway, on food and wine tourism linked to the agricultural and winemaking products of the area, as well as on the natural and environmental beauties or on the historical-archaeological sites, capable of creating and reviving the fortunes of the economy as intrinsic qualities of the nature and history of the places[20].

19 Giorgio and Tenore 1998: 35 et seq.

20 Ibid. p. 42.

Fig. 3. The WWF natural oasis of Conza della Campania with its lake viewed from the train. Source: The author, August 2017.

With its twenty-seven stations, the railway crosses the archaeological areas existing in Irpinia, such as Atripalda and Conza to mention the best known, and the municipalities linked to a strong intangible heritage consisting of folk customs and traditions, such the "Carnival" celebrated as an identity value in Montemarano, Paternopoli and Castelvetere. Other aspects related to the food and wine tourism of DOCG products such as Fiano, Taurasi and Greco wines produced in towns touched by the line, such as Taurasi, Montefalcione and Lapio, as well as other IGP products such as hazelnuts, chestnuts, olive oil, cheeses and cold cuts, should be added to the reasons above. Other aspects are the religious tourism linked to the famous sanctuaries of Goleto, S. Francesco a Folloni or Materdomini; or hiking and countryside tourism in the regional park of the Picentini mountains to admire the sources of the rivers of the region, such as the Calore; or to enjoy the W.W.F. (World Wildlife Fund) natural oasis of Conza della Campania, visible from the railway (Fig. 3). Finally, cultural tourism linked to the history of the individual towns can be added to all these aspects mentioned.

Unfortunately, the wide tourism potential of the area was confronted with the damage caused by the earthquake of 1980. On that occasion a large part of the vernacular architecture as well as almost all the architectural structures serving the line, such as stations, signal boxes and railway warehouses, were destroyed. With the reconstruction of the stations, prefabricated buildings were preferred in most cases. In this way the historical charm of the railway and part of its local characteristics, as an infrastructural work in close interdependence with the territory, have been irreparably lost. Unlike the stations, most of the bridges, both masonry and iron, as well as the tunnels, have only been consolidated, replacing only in some cases damaged structural elements[21]. Among these structures, built according to techniques and materials that denote the desire to create an important railway artery for the provinces of the hinterland, there are nineteen tunnels. The one of Montefalcione is the longest, measuring two and a half kilometres. There are also thirty iron or steel truss bridges[22], including the bridge over the river Calore near Lapio, called "principe", made up of three metal gallery beams of a length of ninety-five metres at a height of more than twenty meters above the ground, supported by only two masonry pillars. Among the masonry and brick works it is worth mentioning the curved bridge over the river Sabato in the town of Atripalda consisting of sixteen brick pillars and spans of eleven metres reaching a maximum height of eighteen metres[23]. Even these works, made with local materials but using foreign technologies, represent a further reason to relaunch tourism for the rediscovery of the industrial past of Southern Italy.

---

21 Ibid. p. 53.

22 According to engineer Crugnola of the *Società Italiana per le strade ferrate del Mediterraneo*, the metal girders of bridges and viaducts could be traced to five types: three types with parallel boards respectively with a cross shape, a simple or double triangular truss and uprights inclined ends, or multiple trusses; again, straight with a semi-parabolic upper board and the ones with parabolic (bow-sting) boards. Ogliari 1975: 892.

23 Pane 2008: 1291 et seq.

Fig. 4. Walk on the railway tracks during the railroad closure, organized by local associations and volunteers. Source: InfoIrpinia and InLocoMotivi, June 2015.

After the definitive closure to passenger service on December 13, 2010, the local associations, first of all "InLocoMotivi"[24], and the various groups of volunteers immediately mobilized for a tourist offer linked to the territory that would revive the railway. They started to re-launch and enhance the territories of Irpinia crossed by the railway, to focus on the landscape values and natural scenarios offered by journeys by train as a new frontier of slow tourism. Nevertheless, after the first initiatives of 2010 in which the train journey was sponsored during some events linked to the territory and organized by the neighboring municipalities, the route was definitively closed to rail traffic. Political decisions, however, which also considered the reuse of the railway as a greenway[25], did not reduced the enthusiasm and commitment of the associations. During the years of closure of the line, events and walks were organized along the tracks to keep the memory of the railway alive (Fig. 4). The associations, the volunteers and the inhabitants of the places have always opposed the idea of the reconversion of the railway track into a greenway, both for the redundancy of a greenway within a territory with a high agricultural and natural vocation, and for the need of infrastructures in the

[24] More information provided at: https://avellinorocchetta.wordpress.com/?s=2010 (August 2018).

[25] Fiore and Sicignano 2015: 70–72.

internal areas without other railway connections, as well as for the excessive length of the route, the longest abandoned railroad in Campania.

A first turning point came in 2016 when the line finally succeeded in attracting the attention of *Ferrovie dello Stato* and in particular of the FS Foundation[26], a national body founded in 2013 following the increasing attention given to railway heritage at a national and international level and the positive feedback, in terms of public attendance, on the occasion of historical trains sporadically organized in Italy. Although the reopening for the local transport was not considered enough advantageous to guarantee profits and the idea of a railroad serving the industries of the internal area was abandoned, it was understood that a rebirth of the line could only take place as a tourist railway, using trains with historical convoys able to attract visitors and tourists in the inland areas.

Fig. 5. A picture from the first historical tourist train. Some passengers are enjoying the natural landscape and the historical engineering works. Source: InfoIrpinia and InLocoMotivi, August 2016.

The first historical train was organized in the summer of 2016, although only for four days and only along the route from Conza-Andretta to

[26] More information provided at: www.fondazionefs.it/content/fondazione/it/it/chi-siamo.html (August 2018).

Foggia[27] (Fig. 5). The train, consisting of a diesel locomotive and historic "centoporte" carriages, represented an incentive to participate in Calitri's Sponz Fest[28]. In that occasion, some means of transport from the other municipalities to some stations of the line were also used to increase participation in the initiative. The following year the opening of this section was repeated in the same occasion and, although the opening had been reduced to only three days, the route was extended to Lioni[29]. Because of the years of suspension of the traffic on the line, it was necessary to make the tracks passable again and to test the civil engineering structures.

Thanks to the increasing success of the initiatives, also in terms of public, it became clear that the Avellino-Rocchetta railway was of a greater importance than only local transport[30]. In parallel with the reopening of the line to slow tourism, the *Soprintendenza ABAP* of Avellino and Salerno, in collaboration with the University of Naples Federico II Department of Architecture[31], was carrying out a study aimed at declaring the cultural interest of the entire track from Avellino to Rocchetta Sant'Antonio on the basis of the article 10 paragraph 3d) of the Legislative Decree 42/2004. This provision, aimed at protecting the entire route for its historical and landscape values, as well as for the tourist enhancement of the territories it crosses, acquires extreme importance as it is the first Italian measure of this kind aimed at an entire railway line. So, in 2017 the Avellino-Rocchetta railway became the first Italian railway to be listed for its historical-cultural as well as its landscape values. The results achieved, thanks to the involvement of numerous actors

27 The train run on August 22, 23, 26 and 27, 2016. More information provided at: www.fsnews.it/cms-file/allegati/fsnews2014/2016_08_19_LOCANDINA_FERROVIA_IRPINIA_2.pdf (August 2018).

28 The SponzFest, conceived and directed by the musician and artist Vinicio Capossela whose parents were natives of Irpinia, was born as a music festival in 2013 and is transformed into an event rooted in the territory where artistic, theatrical and visual performances are combined with music. More information provided at: www.sponzfest.it/2018/ (August 2018).

29 More information provided at: www.fondazionefs.it/content/fondazione/it/it/treni-storici/archivio/2017/8/27/sponzfest-express.html (August 2018).

30 Oppido, Ragozino and Micheletti 2017: 1190–1196.

31 The process for the declaration of cultural interest of the Avellino-Rocchetta railway line, the first example in Italy, came from the collaboration between the Regional Secretariat of the MiBACT and the Department of Architecture of the University of Naples Federico II. Luigi Di Muccio, officer of the Regional Secretariat of the MiBACT, and Andrea Pane, professor of Architectural Conservation and Restoration, represented the scientific coordinators of this initiative.

on a local, regional and national scale, were exhibited during a conference organized in November of the same year and which also saw the involvement of the academic world[32]. The conference was also accompanied by an exhibition with several Master's degree theses produced over the years about this railway, further demonstrating the increasingly global character of this infrastructure.

Conclusions

All these initiatives have spurred more local and regional, but also national, institutions to focus more on the possible re-opening of the railway and especially on slow tourism in the areas it traverses. So the local producers, together with the institutions and associations, have become more aware of being the main actors to sponsor and revalue these places. One of the first strategies adopted was that of having a greater number of historical trains during the year. In November 2017, a train was organized in relation to the local festivities connected to the chestnut harvest[33]; in February 2018, during the Carnival, the train reached the town of Montemarano, to enhance the traditions and celebrations constituting an intangible cultural heritage of the area[34]; and only in May 2018 the entire line from Avellino to Rocchetta S. Antonio[35] was reopened.

The ongoing initiatives[36] that over the years have been multiplied, involving and increasing the value of the territory, the great response of the public in terms of participation in the organized events and the

---

32 On the occasion of the conference, the University Federico II of Naples, the Campania Regional Secretariat of the MiBACT, the regional and provincial institutional bodies, the CNR, the FS Foundation and other associations, took part in the debate about the strategies for the enhancement of this railway. Astrella 2018: 162–163. Verderosa 2018: 164–165. More information provided at: www.fondazionefs.it/content/fondazione/it/it/eventi-museo/archivio/2017/11/17/giornata-di-studi-avellino-rocchetta.html (August 2018).

33 More information provided at: www.fondazionefs.it/content/fondazione/it/it/treni-storici/archivio/2017/11/4/il-treno-delle-castagne.html (August 2018).

34 More information provided at: www.fondazionefs.it/content/fondazione/it/it/treni-storici/archivio/2018/2/11/treno-storico-foggia-montemarano.html (August 2018).

35 More information provided at: www.fondazionefs.it/content/fondazione/it/it/treni-storici/archivio/2018/5/26/treno-storico-benevento-rocchetta.html (August 2018).

36 More information provided at: www.fondazionefs.it/content/fondazione/it/it/treni-storici/archivio/2018/8/5/treno-turistico-benevento-avellino-lapio.html (August 2018).

constant attention of the local bodies involved, as well as of the institutions, allow us to hope for an even more positive future. At the same time the bottom-up strategies, which also pushed other non-railway associations to promote the territory[37], have shown how the recognition of a wide heritage both material and immaterial and the exploitation of an almost isolated internal territory can start as strategies on a local scale and become programmed actions able to create a network among different actors, provoking an echo capable of overcoming local and geographical boundaries.

Bibliography

Astrella, C. I. (2018). Avellino-Rocchetta: Un treno dal passato al futuro. *'Ananke*, 84(5), 162–163.

Battisti, F. M. (Eds.). (2001). *Eugenio Battisti. Archeologia industriale.* Jaca Book.

Bernardini, E. (2003). *In viaggio con i treni d'epoca. 18 itinerari turistici attraverso l'Italia*. Trenitalia.

Conlin, M. V., & Bird, G. R. (2014). Railway heritage and tourism: Themes, issues and trends. In Conlin, M. V. & Bird, G. R. (Eds.), *Railway Heritage and Tourism. Global Perspectives* (pp. 3–16). Channel view publications.

Cuéllar, D. (2016). From railways to heritage: The closure of railway lines in Spain and their valorization as a cultural good. In McCants, A., Beira, E., Lopes Cordeiro, J. M., Laurenço, P. B., & Silveira Pereira, H. (Eds.), *New Uses for Old Railways* (pp. 33–52). CreateSpace Independent Publishing Platform.

Di Marzo, D. (1885). *Le ferrovie di quarta categoria e la provincia di Avellino.* Tulimiero.

Dinhobl, G. (2016). Opening of mountainous and peripheral regions by main and branch railway lines. In McCants, A., Beira, E., Lopes Cordeiro, J. M., Laurenço, P. B., & Silveira Pereira, H. (Eds.), *New Uses for Old Railways* (pp. 121–132). CreateSpace Independent Publishing Platform.

Fiore, P., & Sicignano, E. (2015). La riconversione funzionale delle ferrovie dismesse. Il caso della linea Avellino-Rocchetta Sant'Antonio. In Moccia,

37 The reference goes in particular to the associations most involved such as the Italian Touring Club and InfoIrpinia.

F. D. & Sepe, M. (Eds.), *Urbanistica Informazioni Special Issue. IX INU Study Day. Green and Blue Infrastructures, Virtual, Cultural and Social Networks* (session V, pp. 68–72). INU.

Giorgio, G., & Tenore, A. (Eds.). (1998). *Viaggio nel futuro...e questione meridionale.* Tip. Pannisco.

Giuntini, A. (2001). *Il paese che si muove: Le ferrovie in Italia fra '800 e '900.* Franco Angeli.

Godoli, E., & Cozzi, M. (Eds.). (2004). *Architettura ferroviaria in Italia: Ottocento.* Ed. Dario Flaccovio.

Lenzi, M., Colucci, A., Toni, F., Ricciardi, F., Barra, P., Belli, L., Schirillo, S., & Prezioso, L. (1884). *Ferrovia Avellino-Ponte S. Venere. Petizione della Giunta esecutiva al Ministro dei Lavori Pubblici.* Tulimiero.

Maggi, S. (1997). *In treno per diporto. Dal turismo ferroviario alle ferrovie turistiche. Esperienze e prospettive.* Copinfax.

Maggi, S. (2002). Le role éconimique et social des Chemins de fer secondaires en Italie. *Revue d'histoire des chemins de fer*, 24–25, 370–394.

Maggi, S. (2012). *Le ferrovie.* Il Mulino.

Mainini, G., Rosa, G., & Sajeva, A. (1981). *Archeologia industriale.* La Nuova Italia.

Melisurgo, G. C. A. (1884). *Prima relazione al progetto della ferrovia Trajana Avellino per valle del Calore, valle Ufita, valle Ofanto a Ponte Santa Venere.* Giannini.

Ogliari, F. (1975). *Terra di primati* (vol. III). Cavallotti.

Oppido, S. (2014). La valorizzazione diffusa: Il riuso del patrimonio ferroviario dismesso. *BDC Bollettino del Centro Calza Bini – Università degli Studi di Napoli Federico II, Complex Evaluations for Hybrid Landscapes,* 1(14), 221–235.

Oppido, S., & Ragozino, S. (2015). Linee ferroviarie dismesse. Nuove prospettive di valorizzazione e sviluppo locale. In Moccia, F. D. & Sepe, M. (Eds.), *Urbanistica Informazioni Special Issue. IX INU Study Day. Green and Blue Infrastructures, Virtual, Cultural and Social Networks* (session VII, pp. 32–36). INU.

Oppido, S., Ragozino, S., & Micheletti, S. (2017). Riuso del patrimonio ferroviario (non) dimenticato e processi di rigenerazione. Avellino-Rocchetta Sant'Antonio: Il treno irpino del paesaggio. In Urbanistica è/e azione pubblica, XX SIU National Conference Proceedings, June 12–14 (pp. 1187–1197). Planum Publisher.

Pane, A. (2008). Alle origini dell'ingegneria ferroviaria in Campania: La costruzione della linea Avellino-Ponte S. Venere (1888–1895) e gli attuali problemi di conservazione. In D'Agostino, S. (Eds.), Storia dell'Ingegneria, Atti del 2° Convegno Nazionale (pp. 1291–1300). Cuzzolin.

Rovelli, R., Senes, G., & Fumagalli, N. (2004). *Ferrovie dismesse e greenways*. Associazione Italiana Greenways Onlus.

Stefanovic, K., & Koster, R. (2014). Railfans and railway heritage tourism. In Conlin, M. V. & Bird, G. R. (Eds.), *Railway Heritage and Tourism. Global Perspectives* (pp. 26–41). Channel View Publications.

Verderosa, F. (2018). Avellino-Rocchetta. *'Ananke*, 84(5), 164–165.

Section V

Promotion

Balancing between local and global: Heritage presentations of central European small towns (Case of Telč and Bardejov)

Jan Krajíček

Introduction

Through the case study of the heritage presentations of Telč and Bardejov, this chapter will focus on the strategies of heritage presentation of two specific central-European historical towns included in the UNESCO World Heritage List. Both Telč in the Czech Republic and Bardejov in Slovakia have a long history and deep tradition of cultural heritage represented by many tangible objects as well as by intangible festivals, traditions and historical communities of multiple ethnicities and religions. However, these small towns are located in remote peripheral areas and they are long-term underdeveloped in the terms of socio-economic indicators. Their huge set of cultural heritage is therefore deliberately used for the purpose of the towns' propagation, to attract tourists and businessmen – in short, to promote the towns in a positive way and to improve their condition. The question is, how do the towns operate with this recently acquired UNESCO status, a status of universal importance and global attractiveness, in their heritage presentations? How do they approach the UNESCO and the non-UNESCO heritage? Is the UNESCO label seen as a superior value of these towns, making the non-UNESCO heritage minor, pushed away or even forgotten? How do the towns balance their global and local level of heritage? I will therefore examine the heritage presentations produced by both towns in recent years before and after the UNESCO inscription. These are mostly the printed tourist guides and the official websites. These municipal sources, as well as the claims of municipal representatives, will show the impacts, the advantages and even possible disadvantages and threats of becoming 'a small local town of big global importance'[1].

1. Case of Telč

1.1. Brief characteristics of Telč

Telč is a historical town located on the periphery of the Czech region Vysočina (Moravian-Bohemian Highlands, the hilly and remote rural region halfway between Prague and Brno), in its southern part approximately 30 km far from the border with Austria and near the former border between Bohemia and Moravia within the historical Czech lands. It was established before the fourteenth century, and from the sixteenth century became an important aristocratic manor as well as the regional centre of trade. It was especially well-connected to Austria, but slowly began to decline at the beginning of the twentieth century[2]. Telč has about five and a half thousand inhabitants[3]. Although it is situated alongside one of the main regional motorways, it is quite remote and far from bigger centres, as is the whole rural and hilly region of the Highlands. In long-term perspective, from the first half of the twentieth century onwards, Telč shows some negative indicators of socio-economic development (depopulation, high average age, low education index, unemployment), but this is largely typical for the whole Vysočina region, one of the most underdeveloped regions in the Czech Republic[4].

---

1 For the conceptualization of managing and promoting the cultural heritage as a strategy for (local) modernization in terms of socio-econimic growth, see Urry 1995: 129–170. Urry 2007: 306–339. Cf. Bowitz, Ibenholt 2009: 1–8.

2 For more on the town's history see: Kypta, Beringer, Janoušek, and Rampula 2004.

3 According to the Czech Statistical Office 2018. https://www.czso.cz/csu/czso/pocet-obyvatel-v-obcich-see2a5tx8j.

4 For more on the socio-economical development of the Vysočina region, see: Illner, M. (2001). Regional Development in the Czech Republic efore and after the Accession. In *Informationen zum Raumenentwicklung 11.* Cf. Musil, J., Müller, J. (2006). *Vnitřní periferie České republiky, sociální soudržnost a sociální vyloučení.* Studie CESES 2. Prague.

Fig. 1. The Inner City of Telč, the town's historical centre surrounded by ponds, The Urban Heritage Zone. Source: Archived by the author.

1.2. Cultural heritage of Telč

In 1950 (and again in 1970), the whole historical centre of Telč was declared an Urban Heritage Zone, which is up to today the highest level of urban heritage protection in the Czech Republic[5] (see Fig. 1). The centre around the main square, known as the Inner City, is surrounded by ponds and consists of seven main heritage objects. These are:

1) Gothic fortifications (walls and bastions, ditches, gates and bridges)
2) Gothic Roman Catholic Church of St. James

5 The 1970 Declaration reviewed and broadened the original Urban Heritage Zone complex declared on 11 July 1950. See *Declaration of The Ministry of Culture of Czechoslovak Socialist Republic (27 April 1970).* https://www.kr-vysocina.cz/vismo5/dokumenty2.asp?id_org=450008&id=1010595&p1=1126.

3) Gothic Protestant Church of the Holy Spirit
4) Baroque Jesuit area of monastery, college and Church of the Holy Name
5) Baroque fountains and the Plague Column of the Virgin Mary
6) about 50 Renaissance burgher houses enclosing the square
7) Renaissance Chateau of Telč at the far end of the square.

In 1992, this urban heritage zone of the Inner City was listed on the UNESCO World Heritage List as the first Czech (Czechoslovak) inscription together with Prague and Český Krumlov[6].

The other heritage objects in Telč are remote from the city centre, such as the Jewish synagogue and cemetery, the Baroque Roman Catholic Church and chapel, the water mill and steam mill, and the Museum of Machines and Technology. Telč also has a few heritage sites in the close surroundings of the city, such as the Gothic forest castle Roštejn, and the ruins of the Gothic castle Štamberk. Altogether, there are more than 200 objects declared as various types or levels of heritage, either by the Czech National Heritage Institute or by UNESCO[7].

1.3. Telč heritage presentations

The official website of Telč has two other language versions apart from Czech – German and English[8]. In Czech, the top of the homepage of the web presentation is filled with information useful for local citizens (a catalog of services and office contacts, weather, education, social services, local newspaper). The lower part is devoted to general tourist information (a calendar of the town's cultural activities, opening times, tips for walks). There is no visible mention of UNESCO nor of other heritage objects on the Czech homepage (see Fig. 2).

---

6 UNESCO World Heritage List, *Description of Historic Centre of Telč*. https://whc.unesco.org/en/list/621.

7 Central list of Cultural Heritage of Czech republic. https://www.pamatkovykatalog.cz/.

8 https://www.telc.eu/

Fig. 2. Homepage of the official website of Telč (Czech language version). Source: https://www.telc.eu/.

My comment is that it would appear that in Czech the town does not need to propagate itself through its heritage. Even before the UNESCO inscription, the cultural heritage of Telč, which is traditionally nicknamed 'Moravian Venice', has been so familiar that it needs no further propagation. I will explain this later.

But back to the website, where we switch to German or English – and voilà, the foreign versions clearly differ from the native language (see Fig. 3). The first part of the homepage under the heading 'Tourist Attractions' gives an interactive list of heritage objects and sites in the town and its surroundings. The order of the interesting attractions with their links is: Chateau Telč, Houses on the square (including the House of Telč – a museum and the Firemen's House – a gallery), then Landštejn, Roštejn and Štamberk castles and at the bottom of the list, the town's steam mill and the Technical Museum. Unsurprisingly, this order of objects is very traditional, as we can find it almost completely the same in the tourist guides and leaflets produced from the 1950s until the 1990s, which can be called 'the pre-UNESCO era'[9]. It is apparent that the composition of the heritage list presented by the town did not basically change after the UNESCO inscription. As in the Czech website version, there is no mention of UNESCO in the English or German version, not even under the links which provide otherwise comprehensive information about selected heritage objects. The town of Telč is presenting its heritage as a whole complex, not just picking the most prestigious objects and not distinguishing between the UNESCO and non-UNESCO parts of its heritage.

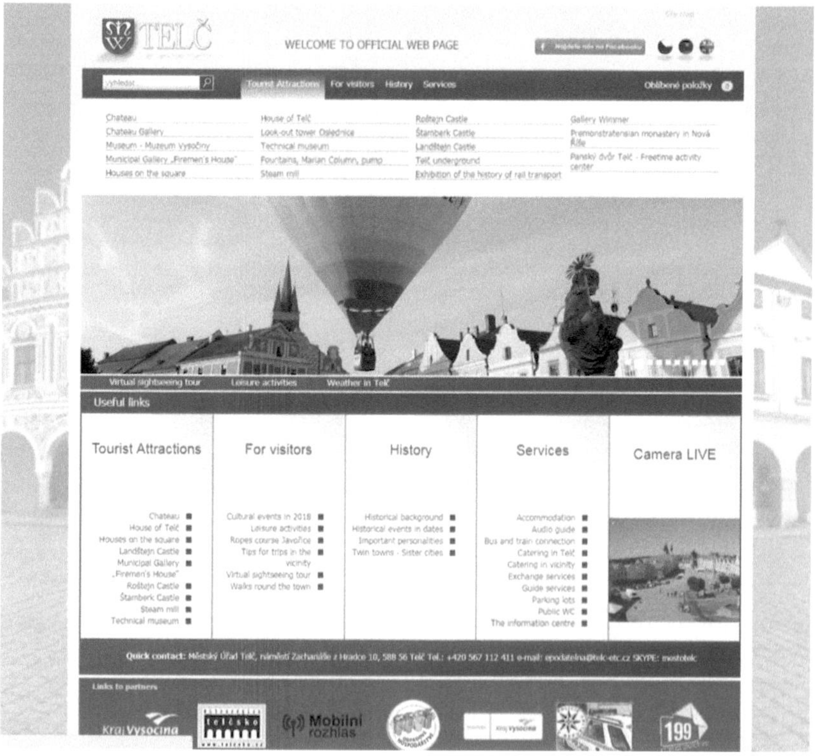

Fig. 3. Homepage of the official website of Telč (English language version). Source: https://www.telc.eu/.

[9] See Kožík 1959. Cf. for example, Kusák 1965, Rampula 1946, and Tiray 1913.

Do the town representatives see the UNESCO label as being so self-propagating and globally-known that the town of Telč doesn't need to support it further?

These remarks and questions were explained in an interview with Roman Fabeš, long-term Mayor of Telč. In Telč, a town of five and a half thousand inhabitants, the volume of visitors per year doubled from 200,000 at the beginning of the 1990s to almost 400,000 nowadays, 35 % of whom are foreigners (mainly from Japan, Germany, Russia and Britain). Mr. Fabeš claimed that even though UNESCO is such a successful global trademark and indisputably the main attraction for visiting Telč, they are not seeking for more visitors, because the capacity of the tourist infrastructure of this small town is already full. Telč is currently looking for better quality, not quantity. Today, the UNESCO-listed centre of the town can be seen in half a day, which is what many foreign tourists are doing. The town's strategy in presenting its heritage is to offer a more complex set of heritage, leisure and sport targets and activities, so the visitors will stay in the town or in the region for more days[10].

The case of Telč, a small peripheral town which has been experiencing the universal advantages of UNESCO inscription for almost 30 years now, shows that the strategy of 'sustainable growth' through the cultural heritage is more complex, more local-based and decentralized, not just simply focusing on the self-promoted and global UNESCO trademark, which is used rather to support and connect the local heritage, not as the one and only topic of the heritage presentation. Overcrowding of the town is seen as a threat, while the dispersion of the UNESCO-attracted visitors to the wider town environs, to local heritage-, sports- or leisure-sites is seen as a development opportunity.

---

10 Information acquired during the international conference '*Culture and Tourism in AT-CZ Region*' (Jihlava, 21 June 2018). Cf. radio-interview Zuzana Kuchyňková, '*Telč má štěstí, že je za jeho nádhernými fasádami taky život* ', Český rozhlas (9 November 2017), <https://www.radio.cz/cz/rubrika/cestujeme/telc-ma-stesti-ze-je-za-jeho-nadhernymi-fasadami-taky-zivot>, accessed 27 August 2018. For general economic analysis of Telč, see Němcová 2014.

2. The case of Bardejov

2.1. Brief characteristics of Bardejov

Bardejov has much in common with Telč. It is also located on the periphery, in north-eastern Slovakia, approximately 20 km from Poland, in the historical region of Šariš (which is also a hilly, remote and rural area). Established in the thirteenth century, it became the crown city and a centre of trade for the whole of Šariš (Sáros), which was one of the wealthiest Hungarian regions. Hence in the seventeenth century, Bardejov began to decay rapidly, for it was struck by plague and almost burned down in a great fire. Until halfway through the twentieth century, Bardejov as well as the whole region was constantly depopulated. Although the period of post-war Socialist construction brought new infrastructure and inhabitants to the town, Bardejov is still difficult to reach by public transport, and seen as a remote and distant place surrounded by the mountains. Even though its population is much larger than that of Telč (cca. thirty thousand), Bardejov experiences very similar characteristics of underdevelopment as the Czech town (depopulation, high average age, unemployment, underdeveloped transport infrastructure), which is also typical for the whole Šariš region[11].

---

[11] For more on town's history see Holotík 1975. Cf. for example Frický 1976, Krpelec 1935. For the socio-economical development of the Prešov region in the 20th/21th centuries, see Matlovič, R., Matlovičová, K. (2005): 66–88.

Fig. 4. The Inner Town of Bardejov, the town's historical centre surrounded by fortification walls, The Urban Heritage Zone. Source: Archived by the author.

2.2. Cultural heritage of Bardejov

Compared with Telč, Bardejov has a very similar set of cultural heritage. The historical centre ('Inner Town'), is located within the fortification walls around the main square. Like Telč, it was declared a Czechoslovak Urban Heritage Zone in 1950[12] (see Fig. 4). The Inner Town consists of four main heritage objects. These are:

1) Gothic fortifications (walls, bastions, and gates)
2) Gothic Roman Catholic Basilica of St. Giles
3) Gothic- Renaissance Town Hall
4) about 30 Gothic and Renaissance burghers' houses.

The Jewish town ('Jewish Suburbia'), consisting of few blocks of houses and a synagogue, is located along the outside of the Inner Town's walls. In 1970, the Old Synagogue was declared a National Cultural

---

[12] As well as Telč, see above Ref. 4.

Heritage building, which is up to today the highest national level of heritage object protection[13].

In 2000, eight years after Telč, Bardejov was also added to the prestigious UNESCO list. The whole complex of the historical centre plus the nearby Jewish Suburbia succeeded in being inscribed on the UNESCO World Heritage List. In the application accepted, the stress was put not just on the picturesque square with the main buildings and surroundings, but also on the multi-ethnical tradition of Bardejov, documented by the Jewish Suburbia close by, whose buildings were in fact very ruined and deserted at the time[14].

The other heritage objects of Bardejov are, as in Telč, quite remote from the centre, such as the Protestant, Orthodox and Greek-Catholic Churches, the well-known historical Spa (Bardejovske kúpele) with the open-air museum of Šariš wooden architecture, and the huge ruin of Zborov castle. Altogether, there are more than 180 objects declared as various types or levels of heritage, either by the Slovak National Heritage Institute or by UNESCO in Bardejov and surrounding environs[15].

2.3. Bardejov heritage presentations

Although Bardejov has a very similar set of cultural heritage to Telč, the town presents it completely differently. In the case of Bardejov, the notion of UNESCO is present almost everywhere (see Fig. 5). All the language versions are the same (Slovak, English, German, Hungarian, Polish, and Russian). The official municipal website places UNESCO in the most visible position on the homepage, and there are many other notions of UNESCO elsewhere in the text. There is a map, and in the list of Bardejov heritage buildings and places with short descriptions, those of UNESCO naturally come first. The municipal website completely omits the heritage located outside the town, such as the famous Spa or the huge Zborov Castle – it is almost completely focused on the

---

13 Declaration of The Ministry of Culture of the Czechoslovak Republic, 11 July 1950. In *Kutnohorsko. Vlastivědný sborník* (2001). Kutná Hora. 58. Cf. above Ref. 4, Ref. 11.

14 UNESCO World Heritage List, Description of Bardejov Town Conservation Reserve. https://whc.unesco.org/en/list/973.

15 Immovable Cultural Heritage Registry of the Slovak Republic. http://www.pamiatky.sk/sk/page/register-nkp-tabulkove-zoznamy.

UNESCO heritage. Moreover, there is just a short mention about the Jewish suburbia, which is part of the UNESCO heritage, but has been also quite ruined for a long period and is still not completely restored even today.

Fig. 5. Homepage of the official website of Bardejov (Slovak and English language version). Source: https://www.bardejov.sk/.

My comment is that it seems as though the municipal website is somehow trying to hide this not-so-picturesque part of the town, and is not willing to propagate buildings that are not under the full management of the town. Is this a strategy, or is it just unintentional?

In fact, the suppression of the Jewish Suburbia had a long tradition in the 'pre-UNESCO era' of Bardejov's tourist presentations. The traditional set of cultural heritage presented by the town consisted mainly of the Inner Town and the Bardejov Spa, with not much mention of Suburbia even after 1970 when the synagogue was declared a National Cultural Heritage building. Nowadays, Suburbia is managed separately by the Slovak Jewish Union, with little cooperation with the representation of the town.

For a complex presentation of the town's heritage, visitors need to follow the link to the Bardejov Tourist Information Office, which is a separate organization run by the town to propagate its heritage (see Fig. 6)[16]. The homepage is again in many language versions (as well as Slovak, in English, Polish, Hungarian, Russian, Spanish, German and Slovenian), and filled with various interactive UNESCO mentions, which are even present on every sub-page. The tourist web offers two main tourist targets – the UNESCO complex and the Spa. In addition, it shows the countryside surroundings of Bardejov (with the Castle and wooden architecture) and offers tourist information (accommodation, catering, leisure and sport activities). The UNESCO-based heritage is displayed at the top, separated in four frames: the walls bordering the square, the Renaissance City Hall, the Basilique of St. Giles and the Jewish Suburbia, which forms the central part of the web presentation.

16 The Office is run by the Municipal Cultural and Tourist Centre (KTC Bardejov), established on 1 July 2008 as te town's public budget organization, cooperating with the local Tourism Office Šariš-Bardejov, organized by the Prešov Region.

Fig. 6. Homepage of the official website of Bardejov Tourist Information Office (Slovak language version). Source: Source: https://www.bardejov.sk/.

Compared with Telč, we can see clearly the different strategy of heritage presentation, and the different approach to the UNESCO-status. It seems as though the heritage labeled with the UNESCO-mark is completely dominant and more important than the other buildings and sites at a lower (national or local) level of heritage protection. While Telč is somehow trying to provide a more complex image of its heritage, Bardejov, on the contrary, is strongly focusing on the UNESCO label, which is absolutely dominant in the current presentation of Bardejov. In the perspective of these findings, I would like to point a difference in the visitor statistics of the two towns. In Telč, about 35 % of visitors come from abroad, so the vast majority comes from the Czech Republic. In Bardejov, 49 % of the visitors are from abroad. According to the survey, most of them are attracted just by the plain fact that Bardejov offers

UNESCO heritage (except for a strong group of Czech visitors, who are mainly attracted by the famous Spa). Again, the strategy is explained by town's representatives. The long-term mayor, Boris Hanuščak (in office from 1999), claims that Bardejov, as a remote and distant city in an underdeveloped region, has no other choice than to exploit the global advantages of the UNESCO mark, whatever it costs.[17] UNESCO is somehow seen as an instrument of 'internationalization', of gaining supra-regional attention. Besides the growth of tourism, which, contrary to Telč, is more focused on foreigners and thus has not reached its limits yet, the general strategy of Bardejov's heritage presentation is to utilize the globally well-known UNESCO mark to attract entrepreneurs and businessmen, and to obtain some 'comparative advantage' in gaining subventions and infrastructure investments on a sub-national level (like highway modernization, and a better connection to Poland). Communications, the connection to Bardejov, is currently seen as the worst obstacle to growth, with UNESCO-status being used as a base for the demands for modernization.

Final remarks and conclusion

Even though both Telč and Bardejov as minor peripheral towns show similar development problems, their approach to the utilization of their cultural heritage differs according to the possibilities and potentials of their growth. In Telč, the tourism strategy is not to raise the numbers of those coming to town, but to decentralize the current number of visitors, successfully attracted by the global, centralist and in the terms of propagation well-functioning UNESCO label, and spread them to other tourist destinations in the close environs of the town so they will support the local infrastructure of tourist services. The presentations of Telč are therefore focused on the wider environs of the town, giving a complex image of a local, not just urban, heritage set. The municipality

17 See the mayor's public interviews on tourism, cultural heritage and urban development. Hudák, M. (2017). Boris Hanuščak: Stále je pred nami dosť roboty. In *Bardejovský reportér 10/10/2017.* https://bardejov.dnes24.sk/boris-hanuscak-stale-je-pred-nami-dost-roboty-281411. Panel Discussion on the International Conference '*Benefits Beyond Inscription: Leveraging the UNESCO Brand for Sustainable Tourism Devolopment*' (Bardejovské Kúpele, 5 May 2017). For reflections and suggestions on the town's touristification of heritage, see Kmeco and Slivová 2009.

presentations do not even explicitly remind one of the UNESCO status, because they considered this so well-known and self-presenting. Bardejov shows a different strategy – more selective, with more emphasis on the global aspect of its heritage, represented by the most valued parts in the central town. Bardejov is still in the stage of tourism growth, with its potential not yet filled, so its strategy of presentation is very municipal-centralized, dominantly based on UNESCO as a universal and international attraction. The municipality representatives are aware of the development problems and infrastructural inadequacy of the town. They use the UNESCO label to overcome the local, peripheral status, to obtain more 'global attention', publicity and support for the sake of improvement to the town's development problems. The heritage presentation of Bardejov is therefore almost dominantly focused on the pivotal UNESCO site, the centre of the town, and does not fully stress other possible tourist attractions in the nearby environs (such as the Spa).

Bibliography

Bowitz, E., & Ibenholt, K. (2009). Economic impacts of cultural heritage – Research and perspectives. *Journal of Cultural Heritage, 10*, 1–8.

Frický, A. (1976). *Bardejov – kultúrne pamiatky*. Bardejov.

Holotík, L. (1975). *Dejiny Bardejova*. Bardejov.

Kmeco, Ľ., & Slivová, A. (2009). Pamiatky jako súčasť produktu cestovného ruchu Bardejove. In *Bardkontakt 2009: Problematika mestských pamiatkových center*. Mesto Bardejov, pp. 64–68 http://www.bardejov.sk/images/stories/o_meste/unesco/problematika_mest_pam_centier/zbornik_prednasok_bardkontakt_2009.pdf.

Kožík, F. (1959). *Až přijdete do Telče*. Telč.

Krpelec, B. (1935). *Bardejov a jeho okolie dávno a dnes*. Bardejov.

Kusák, D. (1965). *Telč*. Prague.

Matlovič, R., & Matlovičová, K. (2005). Vývoj regionálnych disparít na Slovensku a problémy regionálneho rozvoja Prešovského kraja. In Acta Facultatis Studiorum Humanitatis et Naturae Universitatis Prešoviensis, Prírodné vedy, Folia geographica XLII/ 8. PU Prešov.

Němcová, L. (2014). *Analysis of tourism in the town Telč*. University of South Bohemia in České Budějovice. https://theses.cz/id/7nh5x0/BP-Lenka_Nmcov.pdf.

Rampula, J. (1946). *Průvodce Telčí.* Telč.

Tiray, J. (1913). *Vlastivěda moravská II. Místopis Moravy – Díl IV. Jihlavský kraj, Telecký okres.* Brno.

Urry, J. (1995). *Consuming of Places.* London.

Urry, J. (2007). Gazing on history. In Smith, L. (Ed.), *Cultural Heritage. Critcal Concepts in Media and Social Science, Vol. III. Heritage as an Industry* (pp. 306–339). London.

Entry into the UNESCO club: The experience of the Ulyanovsk region

Elena Elts

Introduction

Ulyanovsk strategies in culture were recognized by the international community in 2011 when the city was named Cultural Capital of the Commonwealth of Independent States within the framework of the international program designed to contribute to the expansion of cultural exchange, preservation of historical and cultural heritage and tourism development.

In October 2016, the experiment by Ulyanovsk on development of the territory through culture was presented in the World Report of UNESCO at the Third United Nations Conference on Housing and Sustainable Urban Development (Habitat-III) in Quito, Ecuador. Since the mid-2000s, the city has adopted a cross-cutting approach combining urban heritage and cultural and creative industries. The innovative culture-based strategy has been associated with the dedicated foundation "Ulyanovsk – the Capital of Culture" to manage urban cultural projects. Since its founding in 2012, the foundation has become a lever to develop international cultural cooperation in the Ulyanovsk region.

Forging links with UNESCO has been the ambitious target of the foundation. The intention to submit an application to nominate the city's historic centre for inclusion in the World Heritage List was expressed at a time when the International Cultural Forum in Ulyanovsk was granted UNESCO's patronage in 2013. This was followed by the successful attempt to include Ulyanovsk in UNESCO's list of Creative Cities, which resulted in the use of its branding tool in the framework of local communication strategies and effective cooperation with the cities of the network. It seems realistic to expect that Undoriya (the Ulyanovsk

Region) will be among the first Russian Geoparks to be included in the UNESCO Global Network. The accumulation of experience raises the issue of the possible creation of a Regional Committee for UNESCO in the Ulyanovsk region.

The chapter focuses on the foundation's innovative approach to cultural heritage management in promoting the Ulyanovsk region, and its influence on the cultural policy of the region.

The international activities of the regions of the Russian Federation are rather new. After the break-up of the Soviet Union, an outlet on the international stage was considered by numerous constituent entities as a way to improve their social and economic situation. The legal basis for the international activities of constituent entities of the Russian Federation emerged mainly at the turn of the millennium. The implementation and coordination of international activities at the regional level are carried out by executive authorities of the constituent entities of the Russian Federation: the chief executive who represents a region in international relations makes the decision to negotiate and to sign international agreements, and special bodies (ministries, departments, agencies). Basic forms of the international activities of regions are international agreements, representations of constituent entities abroad, the region's participation in international organizations (cooperation with UN agencies), presentation of the Russian regions in the Press Centre of the Ministry of Foreign Affairs[1]. In the post-Soviet period, the new objectives, practices and technologies of the cultural policy, understood as multi-subject and multi-level, were launched on a parallel track to decentralization, which intensified the process of the distribution of power and responsibility between the federal and the regional level and led to the development of federal and regional programs. The actors of cultural policy at the regional level have gained a particular relevance. The regional cultural policy is generally characterized by autonomy. Heads of regions assume ownership of the initiatives and the regional projects, and support innovative activities[2]. In keeping with the trends hailed by the international community, the regions were beginning to see culture as a socio-economic resource, an impetus to the renewal of regional territorial landscape. At the level of the constituent entities,

1 Busygina, Lebedeva 2008.

2 Astafieva and Avanesova 2015.

a strategy to develop a culture evolves, taking into account the socio-economic characteristic of the development of the Territory. A new insight into the cultural policy can be considered in conjunction with branding territories, one of the key strategies for regional development, which became very fashionable in Russia in 1990–2000[3]. The Russian practice of territorial branding demonstrates a significant emphasis on the preservation of national cultural and natural heritage. Heritage can act as the major factor of branding; for example, inclusion on the World Heritage List turns the territory into an international brand[4].

Territorial branding

The Ulyanovsk region possesses the characteristics of a depressed region and has a subsidized economy. Ulyanovsk, only one of six important towns of the region, is growing in size. Before 1980 Ulyanovsk was one of the most visited tourist places because of its ideological status "Ulyanovsk – Lenin's Homeland", the region benefited from government subsidies. So far, in the list of cultural heritage sites of the Ulyanovsk region[5], almost all the monuments of federal importance (except for the Church of the Resurrection, registered in 2016) had been registered with the Government until 1974. Most of the Ulyanovsk monuments of federal importance are sites associated with Lenin and his family or are archaeological sites.

In 2000, the creation of own brand for the Ulyanovsk region stemmed from the need to unite the local population with regard to transformational agenda, for promotion in the Russian and global market. The Ulyanovsk region became one of the pioneers of branding in Russia. Branding had involved a major package of measures, which included structural changes, legislative initiatives, work on building a dialog with the business community and holding cultural events. City branding efforts changed the public attitude towards the city, and mobilized the different groups to develop the brand. A public search for the meaning of the city had been provoked through communicative grounds[6].

---

3 Ignatieva and Lysenko 2015.

4 Yudin 2014.

5 Information provided at: http://opendata.mkrf.ru/opendata/7705851331-egrkn/#{"version":"5b6310bf7e48557a0899e3d1","tab":"build_map"}.

6 Vizgalov 2011.

A regional brand, developed to all the rules of brand management, was to represent that the region was ready for economic take-off. The idea of "Ulyanovsk – the Aviational Capital of Russia" (2010–2016) played an essential part in developing the city's brand "Ready for Takeoff!". But "the project to develop the brand in compliance with all the technology adopted in classic brand management"[7] did not attract tourist flows to the region.

Under the brand "Ready for Takeoff!" Ulyanovsk participated in the Intergovernmental program "Cultural Capitals of the Commonwealth" in 2011, which had been announced in CIS states as "the Year of Historical and Cultural Heritage". The program was aimed at enhancing cultural cooperation between the CIS member States, the intensification of artistic exchanges, cultural and humanitarian capacity-building of CIS territories and cities, the exchange of experience on cultural policy and economics, and management in the field of culture, promoting the competitiveness of the city and regional brands. Cities that had obtained the status of Cultural Capital of CIS carried out a variety of activities, festivals, congresses and exhibitions[8].

Due to the creation of the cultural heritage committee of the Ulyanovsk region in 2007, it is one of the few regions of Russia, along with Moscow and St. Petersburg, to have extensive experience in the field of protection of objects of cultural heritage, actively realizing powers in the field of regional rule making. The number of cultural heritage sites in the regional heritage register was increased. Ulyanovsk strategies in culture were recognized by the international community in 2011 when the city was named Cultural Capital of the Commonwealth of Independent States within the framework of the international program designed to contribute to the expansion of cultural exchange, preservation of historical and cultural heritage and tourism development. A program featured the first International Cultural forum with the participation of experts of the Council of Europe, UNESCO and the European Parliament. This International Congress "Culture as a Resource of Modernization" was held in Ulyanovsk in 2011, and is well-known both in Russia and abroad as an expert negotiating forum on regional models of cultural policy. According to UNESCO estimates, the Congress in Ulyanovsk,

---

7 Chilikova 2011.

8 Bogoljubova and Nikolaeva 2018.

where the first results of the project UNESCO Culture for Development Indicators were presented, was one of the international forums highlighting the contribution of the culture and creative industries to sustainable development.

Following the Ulyanovsk first international forum, the Ministry of Culture of the Russian Federation has taken decisions to accord priority to the culture of the Ulyanovsk region and include its main projects in the special federal program "The Culture of Russia (2012–2018)". The majority of applications from the region in 2011 featured cultural conservation activities.

The regional and municipal administration continued to promote the new symbolic brand "Ulyanovsk – the Cultural Capital of the Post-Soviet Area"[9]. The Web portal "Ulyanovsk – the Cultural Capital" was launched. The authorities and public and educational institutions strive to create a favorable image of regional culture, using the media in promoting the achievements of local artists and writers. Therefore, in an endeavor to transform regional identity through the use of communication tools the authorities give priority to advancing the idea of the region's cultural development. Only one-third of the population supports this initiative, which could lead to integration with other regions and countries[10].

The new regional cultural policy

The goals and objectives for the cultural policy of the region are represented by two state programs, "Culture in the Ulyanovsk Region 2014–2018" and "Conservation and State Protection of Cultural Heritage, Located on the Territory of the Ulyanovsk Region 2014–2018". The Ministry of Art and Cultural Policy of the region is the governmental body responsible for the implementation of regional cultural policy. The Committee on Cultural Heritage (since 2014 the Department of Cultural Heritage within the Ministry of Art and Cultural Policy) was initiated by the Government of the Ulyanovsk region in 2008 for the implementation of a unified state policy in the field of the conservation,

---

9 Ahmetshina 2015.

10 Shinyaeva and Kaiumova 2014.

use, promotion, and state protection of cultural heritage in the Ulyanovsk region.

The start of the new cultural policy in 2011 was connected with the realization of the project "Ulyanovsk – the Capital of Culture" and its associated foundation. The foundation, launched by the governor of the Ulyanovsk region in 2012 to attract foreign investments, cooperate with experts from Russia and other countries, and create an attractive outward brand of the region, is the planning office for the project. The primary goal of the fund's activities (its functions have tended to expand) is the Territory's development through providing innovative approaches, mechanisms and ideas in cultural policy implementation and the introduction of the practice of modern technologies of cultural management[11].

The presentation of the project "Ulyanovsk – the Capital of Culture" was submitted to the major international organizations: Inter-state Humanitarian Cooperation Fund (ICHF) and the CIS Humanitarian Co-operation Council, in parliamentary hearings of the EU in Brussel (2011) and UNESCO (since 2014).

In 2012, the Russian Federation was engaged in the preparation of the second Council of Europe cultural policy review. The peculiarity of the national report was the incorporation of the regional dimension in the joint (close cooperation of teams of national and international researchers leads to the preparation of report) review report and the combination of a thematic approach with a regional one. The pilot regions being studied were Ulyanovsk, Mari-El and Omsk. The report stressed the fact that the Government of the Ulyanovsk region gave high priority to the development of the cultural sector as a way to improve the general socio-economic situation, quality of life and "investment attractiveness" of the region. State support for culture was constantly increasing (while the state grants system was developing) and the "surplus" principle of funding enabled the accumulation of resources. The report described the project "Ulyanovsk – the Capital of Culture" and its associated foundation as a driving force for the development of a new cultural policy, the task of which was the modernization of the cultural infrastructure, enhancing the quality of life and people's change to the culture. The review demonstrated the successful implementation of the new measures

---

[11] More information provided at Foundation "Ulyanovsk-the Capital of Culture" 2012–2016. http://ulkul.ru/news/ulkul/.

for Ulyanovsk, which have helped the city to play a more central role on the national and international stage. They were: making use of resources accumulated in earlier period for the region's large-scale projects at a national and international level; engaging a lot of partners in the implementation of cultural projects due to the "Ulyanovsk – the Capital of Culture" project; large-scale international festivals and forums accompanied by a varied cultural program; and efforts taken to publicise the cultural events of the region through the mass media of the Russian Federation, CIS countries and Europe[12].

This report and the documents and projects defining the regional cultural policy reflect the policy's role in the rebranding of the Ulyanovsk region. The new status has drawn the notice of UNESCO and the Council of Europe.

The project "Ulyanovsk – the Capital of Culture" called for a Government Programme of Action in the field of culture and possibility of changing cultural environments. The Government of the Ulyanovsk region and governmental bodies are actively involved in the region's cultural policy, based on two pillars, that is, cultivation of the people's creative potential, and long-term regional development through the innovative economic approaches. The modern approach to the cultural policy implies a shift from the sectoral principle of governance to inter-ministerial and inter-sectoral culture development programming. In 2011 the Council for the Promotion of Culture at the Governorship of the Ulyanovsk region was established, which was to identify the priorities for regional cultural policy and mechanisms of their implementation. The Chairman of the Council was the Governor. In 2014 the Council endorsed the Strategy of the cultural policy in the Ulyanovsk region 2030.

The foundation holds various events, including the annual International Cultural Forum (since 2011), aimed at cultural, art and creative industries development and promoting the Ulyanovsk region with the help of culture on the regional, federal and international levels. Since 2012, priority has been given to the formation of the creative industries cluster in the region. The organization had been considered

---

12 More information provided at: https://publicsearch.coe.int/Pages/results.aspx?k=Presentation%20of%20the%20cultural%20policy%20review%20of%20the%20Russian%20Federation%2E%202013.

as an institute for development in the field of culture while also promoting awareness of the region's cultural policy (the Fund advances the research of the cultural policy) and information about the priority project "Ulyanovsk – the Capital of Culture" in the Ulyanovsk region, Russia and abroad, starting in 2014. International cultural cooperation is an integral activity of the foundation, which became the coordinator of international cultural activities in 2015 in the Ulyanovsk region[13]. Further expansion of the functions of the foundation is planned.

Entry into the UNESCO club

The projects in the regions realized jointly with UNESCO and the events of constituent entities, organized at UNESCO in Paris, established an attractive Russian image in the international community. The cooperation with UNESCO has developed within the framework of the National Commission of the Russian Federation for UNESCO and the Permanent Delegation of the Russian Federation to UNESCO (the third structure, the UNESCO Office in Moscow, existed until 2015). Besides, there are regional Committees for UNESCO in a few constituent entities. The presentation of the regions at UNESCO headquarters is an important form of collaboration with the organization, which yields a high return in terms of the promotion of joint projects of constituent entities. The projects under the auspices of UNESCO indicate their high status and the support of the world community. The National Commission of the Russian Federation decides on the admissibility of the request for this status and the Permanent Delegation of the Russian Federation to UNESCO is undertaking relevant work[14].

The region's engagement with UNESCO is considered a priority in the Ulyanovsk region. This international organization is presented as a crucial partner in the reports of the Ministry of Art and Cultural Policy and of the Government of the Ulyanovsk region. The region stands ready to create its Committee for UNESCO. The heritage and cultural policy were also pointed to as a promising area of endeavor. The region is planning to nominate a part of its cultural heritage to the World Heritage List.

---

13 More information provided at Ministry of Arts and Cultural Policy of Ulyanovsk region 2016). www.ulmincult.ru.

14 Budanova 2017.

In September 2012, the International Assembly of Artists "Autumn of Plastov" was held in the Ulyanovsk region under the auspices of UNESCO.

The foundation "Ulyanovsk – the Capital of Culture" strengthens partnership with the foreign partners, especially cooperation with UNESCO, the Commission of the Russian Federation for UNESCO in the framework of the annual International Cultural Forum and the international projects in the sphere of culture. In 2013, the Third International Cultural forum was held in Ulyanovsk under the auspices of UNESCO. Since that time, subsequent annual international cultural forums in Ulyanovsk have achieved the status, which means the international recognition of events and their importance for the development of regional and national culture. The information on this cultural event is regularly publicized on the UNESCO's website.

In May 2014, the presentation of the cultural and educational potential of the Ulyanovsk region was held at the UNESCO headquarters in Paris to present the Annual Ulyanovsk International Cultural forum, regional programs to preserve the cultural heritage of the Ulyanovsk region and develop of the creative economy.

Since 2010, the Ulyanovsk region has been participating in the project "UNESCO associated schools", which has emerged as a priority area with more than 30 member educational institutions in the region. Ecology, environmental protection, study and conservation of natural and cultural heritage, and education on human rights and the rights of the child were the key areas for the action. Since 2017, assistance in the implementation of the project is provided by the UNESCO chair "Titular Languages in Cross-cultural Educational Space" in the Ulyanovsk State Pedagogical University and the educational institutions have obtained a great opportunity to participate in various international cultural, educational and scientific projects. The chair became an additional regional centre to strengthen the communications of the region with UNESCO and other states.

Inclusion of the Ulyanovsk cultural heritage in the UNESCO World List became the ambitious goal of the foundation "Ulyanovsk – the Capital of Culture". The question of the inclusion of the historic centre of Ulyanovsk in the World Cultural Heritage List of UNESCO had been raised in a meeting between the executive director of the foundation Tatyana Ivshina and the Permanent Representative of the Russian

Federation to UNESCO Eleonora Mitrophanova (Chairperson of the World Heritage Committee, 2011–2012) at the UNESCO headquarters in Paris, May 2013. The Ulyanovsk media and the Web portal "Ulyanovsk – the Capital of Culture" highlighted this news. But in the autumn of 2013 during the III International Cultural Forum the Secretary of the Russian National Commission for UNESCO, G. Ordzhonikidze, referred another priority of heritagization, that is, archaeological sites and natural heritage.

Nevertheless, in 2014 the Government of the Ulyanovsk region granted the status of cultural heritage of regional importance to several dozen architectural sites; in 2015 the Ministry of Art and Cultural Policy gave this status to four sites. That had been preceded by active measures to keep the tangible cultural heritage and financing for the state protection of cultural heritage in the Ulyanovsk region.

It appears that there has been a strong desire to make the regional territory receive UNESCO international status. In 2017, within the International Cultural Forum, the initiative (introduced by Governor Sergey Morozov) to make the Regional park "Undoriya" the first one in Russia to receive the UNESCO Global Geopark status was announced.

The foundation "Ulyanovsk – the Capital of Culture" plays a vital role in fostering linkages with UNESCO. It originated as a separate structure which is working to attract external investment, develop relations of cooperation with Russian and the foreign expert community, and build an attractive territorial brand.

In December 2013, the foundation launched the preparation of an application for membership of Ulyanovsk in the UNESCO Creative Cities Network. The initiative group consisting of the representatives of the foundation, the Department of Culture and Leisure of the Ulyanovsk City Administration, the Ministry of Art and Cultural Policy, libraries and creative residents had been preparing the application for two years.

In 2015, Ulyanovsk became a UNESCO City of Literature and was the only city to represent Russia in this field. According to the Governor of the Ulyanovsk region, this status indicates international recognition of the region as a territory that considers culture as one of its top priorities and uses it as a resource for sustainable development.

The foundation has been actively working on the project "Ulyanovsk – UNESCO City of Literature". In 2016, associated with the 250th anniversary of the historian and writer N.M. Karamzin, the Ulyanovsk region has held about 2000 events, dedicated to his personality. So the anniversary has been included in the calendar of UNESCO memorable dates of 2016.

In October 2016, the experiment of Ulyanovsk on development of the territory through culture was presented in the World Report of UNESCO at the Third United Nations Conference on Housing and Sustainable Urban Development (Habitat-III) in Quito, Ecuador. Deputy Head, Olga Mezina in the report "Working with International Partners to Boost the Potential of a UNESCO Creative City" noted the strengthening of region's partnership with foreign countries since 2011 when the city was named "Cultural Capital of the Commonwealth" which culminates in joining the UNESCO Creative Cities Network in 2015 and emphasized the role of the annual International Cultural Forum, covering world cultural trends and best practices, and city's rich cultural heritage opening up immense potential for development in the future[15].

At the turn of the millennium a new reading of cultural policy in the Ulyanovsk region happened as the result of territorial branding. Culture has been one of the development priorities of the Ulyanovsk region since the early 2000s and the significant contribution to the cultural sector has been regarded as a way to improve the general socio-economic situation, quality of life and "investment attractiveness" of the region. Special emphasis has been placed on image project and the preservation of the national cultural and natural heritage.

Conclusion

The successful integration of the region into the international community is the result of a long effort that has required the broad participation of administrative authorities, cultural institutions and the region's population. The leadership in the region (Sergey Morozov has been Governor of the Ulyanovsk region since 2005) has given consistency and coherence to the Ulyanovsk cultural project. The phenomenon of a strong and motivated governor is a key factor in terms of cultural policy development and policy implementation. So the desire to create a successful brand of the Ulyanovsk region contributed to the appearance

15 Mezina 2016.

of the innovative structure, the foundation "Ulyanovsk – the Capital of Culture" which became an important factor in the implementation of cultural policy of the region and in its international cultural cooperation. The initiatives of the foundation in line with governmental efforts aimed at the cooperation with UNESCO in the field of heritage and cultural policy led to the successful implementation in the Ulyanovsk region of a traditional form of collaboration with UNESCO, Ulyanovsk's inclusion in UNESCO's Creative Cities Network and accumulation of experience of interaction.

Bibliography

Ahmetshina, E. R. (2015). Regional'nyj brending kak uslovie konsolidacii regional'nogo sociuma [na primere Ul'janovskoj oblasti [Regional branding as requirement for integration of regional society (using the example of Ulyanovskaya oblast)]. In S. M. Vdovin & V. V. Markin (Eds.), *Regional differentiation and consolidation of social space of Russia. The Fifth Suharev reading: Materials of the All-Russian scientific-practical conference* (pp. 447 – 454). Scientific Center for Social and Economic Monitoring, Saransk (In Russian).

Astafieva, O. N. & Avanesova, O. G. (2015). Cultural policy and national culture: Prospects of the modern Russian strategic vector. *Jaroslavskij pedagogicheskij vestnik* [Yaroslavl Pedagogical Bulletin], 5, 193–201. (In Russian).

Bogoljubova, N. M. & Nikolaeva, Ju. V. (2018). The interstate program cultural capitals of the commonwealth—The model of effective cultural cooperation. *Dialog: Politika, pravo, jekonomika dialogue* [The dialogue: policy, law, economics], 1(8), 91–98. (In Russian).

Budanova, E. (2017). UNESCO and Russian regions in the system of international relations. *Regional'naja jekonomika i upravlenie: jelektronnyj nauchnyj zhurnal* [Regional economy and management: Electronic scientific journal], 4(52). https://eee-region.ru/article/5213/ (In Russian).

Busygina, I. & Lebedeva, E. (2008). Sub``ekty federacii v mezhdunarodnom sotrudnichestve [The constituent entities in international cooperation]. *Analiticheskie zapiski* [Policy briefs], 3(32), 1–31. (In Russian).

Chilikova, L. (2011). Ready for takeoff! *Open Region*, April, 28–31. http://www.ulregion.com/upload/iblock/581/Open%20region_2011_eng.pdf.

Cultural policy of the Russian Federation: national report, report by a European panel of examiners. https://publicsearch.coe.int/Pages/results.aspx?k=Presentation%20of%20the%20cultural%20policy%20review%20of%20the%20Russian%20Federation%2E%202013.

Foundation "Ulyanovsk-the Capital of Culture" (2012–2016). *Accounts of the activities of the foundation «Ulyanovsk – the capital of culture» 2012–2016.* http://ulkul.ru/news/ulkul/.

Ignatieva, O. V. & Lysenko, O. V. (2015). On cultural policies and strategies of construction of territorial image. *Labirint: Zhurnal social'no-gumanitarnyh issledovanij* [Labyrinth. Journal of Philosophy and Social Sciences], 1, 6–16. (In Russian). https://journal-labirint.com/wp-content/uploads/2015/01/Ignatieva.pdf.

Mezina, O. (2016). Working with international partners to boost the potential of a UNESCO creative city. In *Culture, Urban, Future. Global report on culture for sustainable urban development* (77). UNESCO. http://openarchive.icomos.org/1816/1/245999e.pdf.

Ministry of Arts and Cultural Policy of Ulyanovsk region. (2016). Report on the state of culture in Ulyanovsk region in 2015. http://www.ulmincult.ru/.

Shinjaeva, O. V. & Kajumova, L. H. (2014). The dialogue of authority and population in the context of the civil society formation. *Izvestija vysshih uchebnyh zavedenij. Povolzhskij region. Obshhestvennye nauki* [News of higher education institutions. Volga region. Social science], 1(29), 80–89. (In Russian). https://cyberleninka.ru/article/n/dialog-vlasti-i-naseleniya-v-kontekste-formirovaniya-grazhdanskogo-obschestva.

Vizgalov, D. (2011). *Brending goroda* [City branding]. Fund "Institut of city economy". (In Russian).

Yudin, P. E. (2014). Policy of territories' branding and cultural heritage. *Teorija i praktika obshhestvennogo razvitija* [Theory and practice of social development], 12, 109–111. (In Russian). https://cyberleninka.ru/article/n/politika-brendinga-territoriy-i-kulturnoe-nasledie.

Research survey

Touristification of historic urban centres in the Southern Italian Salento region

Katja Maaria Huovinen

Introduction

The argument of this research survey stems from a plan for a cross-sectional study on the touristification of the historic urban cores of Lecce, Otranto, Gallipoli and Nardò in the Salento geographic region in Southern Italy.

Tourism is transforming the economy, significance, property ownership and the use of urban space in the aforementioned towns. The weak local economy, which shifted from agriculture to an inflated public sector during the 1980s, has made these cities extremely vulnerable and dependent on every kind of investments.

The main driver of change in the process of touristification is the global megatrend of constantly growing tourism, together with the lack of a local economy. The internet has made the branding of a formerly unknown region easier, and short-term rental online possible.

The main problem caused by touristification is rising property prices and rents. The tourists are replacing the inhabitants but only marginally contribute to the local community, and the nightlife quite often consists of shouting, littering and urinating. The proliferation of tourist-oriented services and the standardisation of aesthetics may turn the authentic urban environment into a non-place which will lose its attractiveness and authenticity.

This research aims to evaluate the sustainability of the phenomenon, to identify the right policies to control it, and to avoid a negative outcome of transformation due to the uncontrolled touristification of the historical centre.

Both quantitative and qualitative data are used. The research will study the development of property prices per square metre in the historical centre, compared with the surrounding city, and the official data of registered accommodation facilities such as hotels, rooms-to-rent and B&B's [Bed and Breakfast] compared with the numbers given by various booking sites.

The qualitative data consists of written sources such as the discourse in the local newspapers, on visual sources such as the representation of the "Salento" brand in publicity, and on travel sites. Additionally, the standardisation of aesthetics under the key of "otherness" and exoticisation is labelled "Salento style" in interior design magazines, Pinterest and social media. Interviews with the new local community created by lifestyle migrators forms an oral source for the ongoing transformation.

This research is related to studies of gentrification and touristification, studies of colonialism, and studies on sustainable tourism and cultural tourism. I am eager to focus on the less famous towns, each of them full of hope to transform their uniqueness and beauty into economic gain.

In this text I introduce a research plan and some initial data for my new cross-sectional study on the on-going transformation process of touristification (tourism & gentrification) of the historic urban cores of Lecce, Otranto, Gallipoli and Nardò in the Salento geographic region in Southern Italy. My research will concentrate on the change during the last two decades. The argument has been recently studied in many famous tourist destinations, but from my point of view, the economic, socio-cultural, and environmental impact of tourism's constant search for new destinations "off the beaten track" may be even more devastating for small and formerly unknown locations. I hope my research can give results that are useful for local authorities in the creation of sustainable tourism policies and minimising negative impacts, and for the communities as tools for their responses and resistances, when needed.

Salento

Not corresponding with any administrative division but more with linguistic one[1], Salento has been branded with its name as an entity of its

1 Salento is a sub-peninsula of the Italian Peninsula, which encompasses the entire administrative area of the province of Lecce, a large part of the province of Brindisi

own during the last twenty years. One exemplary milestone for the creation of a public identity in the collective imagination is the first edition of the Night of Tarantula music festival in 1998.

The most important among the historic cities of Salento to be studied is baroque Lecce, the capital of the homonymous province, with its 95,067 inhabitants, nicknamed "The Florence of the South". The tiny seashore town of Otranto with its 5,740 inhabitants on the Adriatic shore and the city of Gallipoli of 20,766 inhabitants on the Ionian shore are attractive for beach tourism. Nardò is an inland town and with its 31,564 inhabitants is the second-largest in the province by population. Its beautiful historic centre happens to have been my hometown for the last twenty years.

The four towns studied differ not only in size, but also in their other qualities. Lecce as the capital of a province has more public sector employment than the others. Besides that, the city has a university, and the student population – both resident and commuter – is a positive influence on the city's livelihood all the year round. Tourism has played an integral role in the transformation of Lecce's formerly poor and partly decayed historic core. Public opinion on the development of tourism in the historic centre of Lecce has assumed quite negative tones during the last years, and controversy has also broken out this season.

The numbers of tourism in Nardò are well below those in other towns, which is not necessarily a negative thing. The city centre has still numerous empty stores and workshops, which during the last years have been transforming themselves one by one into bars and restaurants. Other kinds of activities can hardly reach the point of viability.

Otranto and Gallipoli differ from other towns studied because of their seashore location and the legacy of mass tourism of past decades. Both of them have their tiny historic core full of seasonal souvenir shops. During the last years Gallipoli, often nicknamed "Ibiza of Salento" has been described in the local headlines as an example not to follow, and the registered drop of visitors has been received with glee.

---

and part of that of Taranto. The Salentino dialect spoken in the region is a dialect of the Sicilian language. The region also includes linguistic minorities of Greek and Albanian origin.

The drivers of change

The main driver of change in the process of touristification is the global megatrend of constantly growing tourism and especially urban tourism, together with the lack of a local economy. The decrease in the demand for a labour force demand in agriculture is a constant from the 1950s; the Salento region has never been largely a part of industrialisation, and the few local industries suffer from dislocation and foreign competition. The public sector cuts of the last period have halved the local occupation rate.

The local politicians remember their picturesque historic city at least in electoral speeches and are eager to see it transformed into "the goose that lays the golden egg", especially when it has been more or less abandoned for decades. The new trends and brands, like growing tourism, *movida*, peer-to-peer short term letting and other forms of short term rentals, vacation homes and lifestyle migration are warmly welcomed.

Official speeches paint a picture of cultural tourism, but the multiple stakeholders trying to attract the masses and get their livelihood out of visitors often achieve something that somehow resembles classic Mediterranean mass tourism offering sun, sea and sand (but also sex, spirits and smoke), the noisy Ape Calessino auto rickshaw, fewer museums, more food, drink and entertainment.

The numbers and the economic importance of tourism are growing in the context of the local economy, which during the 1980s shifted its core from agriculture to an inflated public sector. According to my hypothesis, tourism is transforming the economy, significance, property ownership and the use of urban space in the aforementioned cities, as the weak local economy has made the cities extremely vulnerable and dependent on various investments regardless of the consequences.

The internet economy has made the branding of a formerly unknown region much easier. Almost anyone with basic web skills and nice photos can promote a proper business. Among new kinds of opportunities, short-term rental online services are worth a mention.

Problematics

The increasing attention – not only from the local community, but also from foreigners and wealthy Northern Italians – manifests itself in rising property prices and rents. The augmented value might be a positive phenomenon, but at the same time the increasing *expectations* of property value may contribute to long-term abandonment, especially when keeping property empty is not heavily taxed and the property owner doesn't have a real urge to monetise. In the same way, increasing expectations of rental income drive away activities with a low profit margin, and leave space to bars, fast food restaurants, pubs, and – maybe – to some franchising stores.

The new sharing economy phenomenon of short term rentals and the "airification" of the cities is studied in solidified tourist destinations, but not so much in new, possibly even more vulnerable ones. Short term rental brings economic advantages over long-term rental. The tourists are replacing inhabitants, but only marginally contributing to the local community. At the same time, souvenir stores, bars and cafés are replacing grocery stores. The menu prices are rising above the financial means of local inhabitants, following the logic of short-span cream skimming. The most profitable activity seems to be nightlife and *movida*, which quite often create collateral damage, such as littering and urinating, shouting and voicing, high volume music. With the proliferation of tourist-oriented services and chain stores, pizza slices and kebab bars, and through the standardisation of aesthetics, the authentic urban environment may turn into a non-place and lose its attractiveness and authenticity. Urban conservation policy and legislation concentrates mostly on material authenticity, and as there is not a great deal of monitoring other than of monuments, not so much even on that.

The research questions

The main question of my research is: can the actual tourism in Salento be defined as sustainable development, and if not, can it be turned into such? In what ways are the local administrations trying to control the phenomenon? Is it possible to control it without violating the right to property? Is there a property bubble, and what will happen to a "seduced and abandoned" city when the bubble bursts? Presuming the inequality of this economic transformation of the city, who reaps the profits and who bears the burden of a fast and somehow uncontrolled touristification of the historic centre?

Methodology and data sources

My research is mixed; both quantitative and qualitative data are used.

Some of the most important quantitative data sources for the research are real estate agencies, which can give detailed data on the rise of property prices per square metre in the historic centre, compared with the surrounding city. In all locations I will compare the official data of registered accommodation facilities, such as hotels, rooms-to-rent and B&B's to the numbers given by various booking sites, such as Airbnb, booking.com, and note the eventual discrepancy between them. In an observation span of a couple of years I am hopeful of tracing a growing or decreasing trend in figures. Another set of quantitative data traceable from the municipal authorities is the number of activities, followed by an analysis of which of them are tourism-oriented, like souvenir stores, bars and cafés and high season stores.

The number of vacation homes would be important information, if available. Some properties fall into more than one category, being both vacation homes and short-term-rented property. This has led to the creation of a new profession of "airbnb manager", a person available for running errands, handing property keys, calling the "cleaning lady" or a plumber when the "real" host is not in the city.

The Regional Tourism Agency "Pugliapromozione" was established in 2011.[2] It publishes on its site statistics for arrivals and overnight stays for every municipality the region between 2005 and 2017, and, from 2011 on, has reported on tourist mobility for the whole region month by month and by nation of origin, without distinction between municipalities. Between 2005 and 2017 the number of visitors and overnight stays has grown in all four, cases but in the seashore localities Otranto and Gallipoli still outnumbers the bigger inland cities. The growing numbers of visitors and overnight stays do not refer only to independent and DIY-tourism in cultural locations like historical city centres, but also to the traditional kind of mass tourism in bathing localities, big hotels, villages and resorts.

---

2 "L'Agenzia Regionale del Turismo, denominata Pugliapromozione, istituita con Decreto del Presidente della Giunta della Regione Puglia n. 176 del 2011 (come previsto dalla Legge Regionale n.1 del 2011, modificata dalla L.R. 18/2010) è un Ente Pubblico non economico, un organismo tecnico operativo della Regione." https://www.agenziapugliapromozione.it/portal/chi-siamo.

Since I am in the initial phase of my research I still do not have exact data to prove that the number of real estate agencies is growing, but in my local context I have seen many technical and building companies transforming their businesses in that direction. So far it is a hypothesis to be proven.

In the future I will study methodically the variation of real estate prices per square metre in different zones of the town, but in my work as an architect I have seen in the historic centre of Nardò some signs of an over-heated situation in the real estate market compared with the rest of the town. Foreign and Northern Italian clients seek apartments in the historic centre, with a garden or a roof terrace at the same level as the rest of the house, with traditional vaults, and are not interested at all in the "normal housing" outside the city walls. Neither are the local residents, as there are already many empty houses, built for a son who moved to North Italy for work, or inherited from grandparents.

The method for the study of qualitative data is discourse analysis. An important written source consists of the discourse in the local newspapers. In the future I intend to see the archives for Nuovo Quotidiano di Puglia and Gazzetta del Mezzogiorno for the years 1998–2018. A lot has been written, especially about Lecce, and not in positive tones. Lately Gallipoli too has had its fair share of criticism.

The lifestyle migrants, who differ from "expats" on secondment or from immigrants trying to integrate into the local work force, have formed a new local community. Most of them are members of the creative class, "downshifters", teleworking, living off savings, retired or on early retirement, and can take advantage of differences in housing prices or salaries between their new town and their place of origin.

Besides an analysis of how "good and bad tourism" is being treated in local media, I am interested in the birth of a "Salento" brand, and in the way it has been represented in publicity and on travel sites. In interior design magazines, or in Pinterest and the like, it is possible to see a standardisation of aesthetics under the label "Salento style", and also some illustrated books in dual-language editions are being published by a local editor. Some recurrent elements are whitewashed houses, wooden ladders, Indian fig opuntia plants, mangers and washbasins made of local limestone, terracotta vases, straw chairs, local stone vaults and hanging tomato clusters. Paradoxically, these elements are used more for decor and furnishing for tourists and newcomers, as the natives are more inclined to use their terraces and courtyards for conserving firewood, a vespa, or a motor hoe.

A possible oral source for the ongoing transformation are interviews with new inhabitants and their communities.

Related work

Research fields related to mine are studies of gentrification and touristification; studies of colonialism, as the new inhabitants have invested in the city and thus feel entitled to tell the natives how to live in their city; studies of "orientalism", as the branding of the product Salento is partly based on its "otherness" and exoticisation, from Moorish Revival villas to Greek pub names; another rich source of related studies is the field of studies on sustainable tourism and cultural tourism.

Some numbers and images to start

To test my methods, I made a booking simulation on 17 July 2018 on booking.com website: I searched for a room for two persons for the period 22–28 July 2018.

The website offered some zone filters, but a zone filter which made possible to limit the booking search into the historic centre (Old Town) existed only for Lecce. For all the other cities there is a "Guests' favourite area" filter, and for Gallipoli also "Baia Verde" filter. For Gallipoli and Otranto only the site gave an alert of being 78 % reserved for the selected dates.

I made further booking simulations for two nights around the Ferragosto festivity, for a 7-night stay including Ferragosto, for one night, two nights and a whole week in mid-September and a week in mid-October. All the cities showed a difference in availability for single-night bookings and two-night bookings, but for Lecce the difference between two-nights and a week was minimum. In Lecce, the number of properties grew between September and October, in all others it decreased. The peak number of properties in Gallipoli exceeded Lecce, even if the number of inhabitants is less than a quarter of provincial capital's population. Other remarkable differences between Lecce and the other towns was that Lecce still had quite high availability for the period around the Ferragosto festivity.

Another good data source would be the airbnb.com site, and I will utilise it in the future, probably with a data scraping tool.

To get an idea of the Salento region in the collective imagination I used the "Google method" doing a Google image search (25 July 2018) and analysing the content of the first page. The image search with word Salento Google Image Search gave me some maps, but most photos were of the turquoise sea, seashore, cliffs and beach, natural pools. The words "Salento style" gave a page full of local limestone, whitewashed walls, white linen clothing and terra-cotta vases. The word "Lecce" gave some seashore photos (also Gallipoli), one photograph with *trulli* and one with a view of Matera [sic]; most photographs were from the city's historic centre, and the attractions amphitheatre, the cathedral, S. Oronzo Square. The most widespread images of Nardò were the Piazza A. Salandra, the castle, and some seashore photos, but there were also some aerial photographs of the Nardò Ring, a high speed test track located more than 20 kilometres north-west of the town. The images of Gallipoli were mostly air/drone photographs of the historic town surrounded by the sea, and the ones of Otranto are seashore views of town: some photographs portray cliffs and natural pools, and of monuments, there was one photograph of the cathedral and two of the castle.

I repeated the search in Instagram with hashtags (30/07/2018). The most quoted hashtag is #salento with 3,041,598 hits, followed by #lecce with 1,102,140 and #gallipoli with 844,677. The hashtag #otranto was used 314,129 times, and it was mostly associated with sea views and seafood, unlike Gallipoli portrayed ladies in bikini. The numbers dropped for #salentolife used 113,977 times. #nardo without accent gave 59,796 hits, mostly of cars and possibly for the test drive ring. #nardò was used 33,230 times, and #salentostyle only 11,463 times.

Conclusion

Heritage characterization and strategies for resilient small towns

Blanca Del Espino Hidalgo – Luďa Klusáková

This book consists of fourteen chapters, varied in origin and characterization, although they share certain basic characteristics. These, related to heritage in small towns and to heritage/objective strategies and resilience, will allow us to draw some conclusions, either by affinity or by the contrast between the different cases and the results obtained by their authors.

The characterization of small towns

We will start by making a brief assessment of the characterization of the chapters comprising the book: in general terms, they all have in common the fact they work with a case study related to heritage commodification and the resilience of small towns. However, the orientation each one provides on the subject is different, and this is the motivation for structuring the book into sections. It begins with two works that pose general reflections from case studies and are followed by four that speak of projects and strategies for resilience in specific cases. There are then two contributions that put a special emphasis on community resilience, then three more chapters on initiative applied in urban networks and territorial cooperation, and finally two that abound in the international promotion of the heritage resources of small towns, having UNESCO as the main promoter. Finally, it includes a research survey about a case dealing with touristification.

The cases chosen are also diverse concerning their characterization, and not only due to their location, even though they all have in common the fact that they are small towns. For example, the fundamental character of these towns being rural or urban has profoundly marked the type of strategies that have led to heritage, and to the results of these approaches.

In this sense, we have towns in the deep countryside (Vysoké nad Jizerou, Telč, and Bardejov, among others), but also the case of a town in the metropolitan area of a capital city (Viterbo, near Rome). On the other hand, the scales of research have been diverse: in many cases, the scale has been local, but there are also studies on a regional scale, and even making international comparisons, as in Horeczki, Pellegrini and Krajíček.

Likewise, the studies and the processes of objectification are very much conditioned by the heritage character that predominates in each of the cases. The book shows a variety in which, first, a distinction is made between rural heritage, which is generally more focused on nature, landscape and agricultural resources, and urban heritage, in which monuments, relevant historical buildings and fortified sites, among other examples, stand out. In many cases, archaeological vestiges exist, although they do not generally represent the fundamental asset on which heritage strategies are based. In some cases, the tangible heritage gains importance for small places, but we can see also cases that have a marked link with the intangible heritage, normally bonded to local production and organization of work/life issues. We also have cases that focus on industrial heritage, both in the factories themselves and in the transport infrastructure, and on the intangible heritage, through the assessment of specific aspects of the collective memory or the heritage and cultural identity of the inhabitants.

In this sense, we can conclude that heritage resources in small towns are as diverse as in any other urban scale and that the initiatives carried out for the regeneration, promotion or resilience of these places and their inhabitants have a potential independent of the heritage characterization, the scale of work or the degree of urbanization.

Heritagization and commodification strategies

The chapter by Gonçalves et al. begins by indicating that resilience occurs when conditions change and that this change can be gradual (progressive) or punctual (accidental). It makes sense to examine the triggers of the issue for which resilience has been sought through heritage-based strategies. In most of the papers in this book, this is due to a gradual issue that has ended up changing the conditions on which the small towns case studies based their balance and daily life: changes in political conditions, as in the case of Vysoké nad Jizerou and Ulyanovsk, problems of territorial articulation that are alien to the structure of the population and the way it actually functions, as in the Hungarian and South Transdanubian small towns, and problems of the widespread obsolescence of historical centres, as in the case of Viterbo or Salento – although in the last case the most recent conflict is actually due to over-tourism. The mass arrival of foreigners is also a problem in the BioComechingones Corridor, linked to real estate pressure and loss of local identity. In several cases, this is accompanied by conditions of territorial marginality that have been aggravated by the economic crisis, as is the case of Telč and Bardejov, and somehow in Friuli Venezia Giulia, even if a big catastrophe such as an earthquake provoked the first and deeper effect on the small towns' resilience. A common problem is also the disappearance or reduction of the productive conditions of the environment, such as the productive obsolescence of the cascareros in the Guadalhorce Valley in Malaga, the decline of local industry in Naoussa due to economic globalization, the deindustrialization of Avesta and Söderhamn or, in the Italian case, the closure of the Avelino-Rocchetta passenger train line, also greatly affected by an earthquake in the 1980s. The cases in which the detonator is accidental are linked to dramatic events for the population, such as the abovementioned earthquakes in Friuli and Avelino-Rocchetta but, overall, the devastation caused by a large forest fire in Cernache do Bonjardim or the construction of a nearby dam that forced the town of Granadilla to be abandoned has the same effect.

Just as interesting as the origin of the problem that generates the need for heritage-based initiatives for resilience, is the nature or origin of the actions carried out in these small towns. In this sense, the vast majority of the case studies analysed in this volume are based on top-down initiatives, generally promoted by the administrations at different scales, whether national, regional, local, or a combination of the above. Exceptions to the generality are the cases of Cernache do Bonjardim and Vysoké nad Jizerou with a clear grass roots approach in which citizens have played a leading role almost exclusively in the search for community resilience, and to a certain extent in the case of the Avelino-Rocchetta railway line, where initiatives promoted by local associations were eventually replaced by institutional action, or the case of Salento, where promotion occurred in both directions without a clear trend.

This is closely related to the origin of the investment that has necessarily accompanied the different actions carried out. In general terms, top-down initiatives have been accompanied by public investment models, and those that are bottom-up by private investment, but not exclusively. In many cases, moreover, private investment, is fundamentally related to the services necessary for the development of tourism in different places, for example, in Vysoké nad Jizerou and Salento. In some cases, a part of the public initiatives has been aimed at encouraging private investment, such as in Viterbo, where they were invited to rehabilitate and use vacant buildings, as happened in Naoussa, or in the cases of Telč and Bardejov. On the other hand, there are cases where this investment is informal and comes up spontaneously among the population, as in Cernache do Bonjardim.

In some places, investment has been based on inter-institutional partnerships, such as in Ulyanovsk, where a foundation was created that includes non-governmental associations and other institutions. Such alliances also occur in the public sector, where there is much searching for funding through European calls (as in the Guadalhorce Valley or on the border of the Prepotto area with Slovenia), or through the establishment of agreements and alliances between different sectors, or even collaboration with academic institutions as in Viterbo and the Stella river valley, or the creation of alliances and foundations of their own, as in the Avelino-Rocchetta railway line and the aforementioned Ulyanovsk. We have had the opportunity to see the transfer of European models beyond, as in the Argentinian case of BioComechingones Corridor, including the town networking and landscape heritagization strategies.

However, both private investment and the promotion of public investment have necessarily been accompanied by the incorporation of regulations, official guidelines and even urban and territorial planning documents that have made it possible and oriented the generation of actions to heritage commodification. Thus, there are few case studies without clear guidelines from the administrations, planning interventions and even a comprehensive plan at both local and territorial levels. Among the exceptions, we can mention the cases of Naoussa, where they were limited to some early actions after the war; of Cernache do Bonjardim, where a formal and coordinated aid strategy was lacking after the fire; of Vysoké nad Jizerou, where this very scarcity of regulations favored real estate investment, whose excess the local population and the authorities resisted; and of Salento, where local efforts were mainly focused on tourism promotion.

On the other hand, we find cases that had integral plans, such as the strategic plan of the Guadalhorce Valley, the master plan of the historic town of Viterbo, the program for the recovery and educational use of abandoned settlements in Spain that supported Granadilla, the landscape strategy for the areas of the Stella River and Prepotto, the official establishment of the BioComechingones Corridor and its Tourism Plan, and the declaration of tourist and cultural interest of the Avelino-Rocchetta railway line. Likewise, various specific initiatives in terms of planning and regulations had their effects in other cases, such as the plans for the restructuring and the urban renewal of Avesta and Söderhamn, the continuous efforts in territorial and urban planning, trying to reinforce the urban entity of the Hungarian and South Transdanubian small towns, and the new regional cultural policy, added to the two statal plans experienced in the case of Ulyanovsk.

One of the most differential issues in the characterization of the cases has been the importance of local society in the strategies, the occasions on which the actions have been characterized as grass roots, despite their importance in the construction of the concept of resilience, being few. Thus, even in small towns where the initiative has been originated fundamentally by official institutions, community building has not generally been one of the fundamental objectives pursued, or its effect has not been relevant. As exceptions, we can mention the case of the Guadalhorce Valley, where this has been attempted through an educational program, of Granadilla, where the lack of specific programs has been mitigated by the sense of community present in the former inhabitants displaced from their homes, and in a certain way in the case of Ulyanovsk through cultural programs for the local population. In the two cases identified in this book as being based on the local community, this has on the contrary, naturally been one of the main assets of the strategies. In Cernache de Bonjardim it is considered the main strength in the face of adversity, based on the sense of belonging to the place and the territory, while in Vysoké nad Jizerou the self-identification of ethnic singularities and attachment to the territory have been considered the main strengths.

On several occasions, the build-up of the initiatives has been related to the promotion and establishment of educational programmes. We have already mentioned the case of the Guadalhorce Valley, where they constituted a central aspect of the program for the recovery of the heritage of the cascareros and were aimed at different age groups. The cases in which these programs are aimed at younger age groups, however, are not numerous. They include the summer schools in Granadilla, which have also helped to rebuild some of the town's houses, and the participation of the town of Ulyanovsk and its schools in the "UNESCO associated schools" project, enabling young people to take part in the local strategy. On the other hand, several of the cases included in this book refer to the participation of higher education in heritage recovery programs and for the design and establishment of strategies. This is the case of the celebration of an international design workshop for the master plan of the historical centre of Viterbo, the establishment of university centres in the vacant industrial buildings of Naoussa, the promotion of research grants for the landscape strategy project in the Stella river valley, and the existence of research projects and final university master projects on the Avelino-Rocchetta line.

Concerning the definition of strategies, the last characteristic that we will analyze jointly in these conclusions will be the promotion carried out on the heritage of the small towns chosen for the research by each author, since it was considered important given the results obtained in each case. We will begin by analysing internal promotion, that is, that which is aimed at the local citizenry to improve its anchorage in the place or its identification with the town's heritage. In the cases of Avesta and Söderhamn, efforts were initially focused on promoting local investment in industry, later extending to shops and cultural centres, while in the Hungarian and South Transdanubian small towns special promotion was made for population growth, although this was not followed by other necessary measures, such as increasing public facilities. In Viterbo, the sustainable use of the historical centre was promoted, restoring the habitability of empty buildings for housing and the service sector, as was the case with the vacant industries in Naoussa. In Granadilla this strategy was not possible as it was an abandoned settlement but remained the private legacy of the former inhabitants. In Cernache do Bonjardim, despite the importance of the local community in resilience, the absence of an internal promotion strategy promoted by official sources has been evident. Similarly, in Vysoké nad Jizerou, even though the authorities have not promoted it recently, the outcome of past efforts in this direction remains. In the cases of the BioComechingones Corridor and Friuli Venezia Giulia, internal promotion may, if anything, be due to collaboration between the different municipalities following the signing of common agreements or in local cooperation between the different heritage stakeholders, what can also be understood in the Avelino-Rocchetta line, if we refer to local associations in defense of heritage. In cases destined for international projection, internal promotion has not been a priority, except perhaps for the effects of the cultural program within Ulyanovsk.

External promotion has also been a strong pillar for many of the initiatives, as in the case of Avesta and Södernamn, linked to the creation of new commercial and trade nodes, as well as the promotion of tourism. In more rural cases, such as in the Guadalhorce Valley, the promotion of tourism abroad has been formalized in the establishment of the agricultural park and the opening of natural attractions such as the Caminito del Rey. Tourism has not been so relevant in Viterbo or Granadilla, although there has been some promotion abroad linked to educational programs, in the cases of Viterbo in the international workshop, and Granadilla through the summer schools. In places where the sense of local identity is particularly strong, tourism has been discreetly promoted, as in Cernache do Bonjardim and to a certain extent in Vysoké nad Jizerou, where despite the promotion of snow tourism and tourism based on strong local identity, the local community has not felt involved in the establishment of the UNESCO cultural park. In other places, on the contrary, both national and international tourism has been one of the main objectives of the strategies developed. This is undoubtedly the case in Salento, where international and national promotion has become a problem, and to a lesser extent in the BioComechingones Corridor, more oriented towards national tourism, in Friuli Venezia Giulia, related to heritage, natural assets and gastronomy, and in Avelino-Rocchettta, where it has displaced local promotion. Finally, the case studies focused on the search for promotion through UNESCO, that is, Telč, Bardejov and Ulyanovsk, have been focused on this objective.

Heritage-based resilience results

The set of heritage-based strategies, initiatives and actions that have generally consisted on the commodification of the heritage assets of small towns chosen as case studies have led to certain results that, in this book, have been evaluated by focusing on resilience.

We will begin by analysing the degree of success achieved by each of the projects described in the different case studies, always from the diverse authors, whose reasons and details in this respect have been set out in each of the chapters. A good number of authors are optimistic and reveal a high or very high degree of success for their case studies. This is the case of Royo on the initiative carried out in the Guadalhorce Valley, although we must take into account that the project is still in progress, which is also a consideration for Ostos and Aladro on Granadilla, whose pace of development is even slower. Gonçalves, Rosendahl and Pérez are optimistic about the future in Cernache do Bonjardim, as they are confident in the capacity of social resilience to face new difficulties. Janáč also makes a positive assessment of the experience in Vysoké nad Jizerou, since most of the activities, as well as the population, have been maintained in this small town. A similar thing happened in Ulyanovsk, according to Elts. Astrella has observed an increase in success from the time of the opening of the project in Avelino-Rocchetta up until today. Krajíček observes optimistic results in his two case studies, Telč and Bardejov, with probably too high a success for Telč, which could threaten the local balance. The success of the initiatives for the promotion of tourism in Salento also seems excessive as observed by Huovinen. Other results are not considered as successful by the authors responsible for these case studies. Nilsson is sceptical about the results achieved in Avesta and Söderhamn, even though the population is starting to grow and renew itself, while Horeczki is generally pessimistic about the low degree of resilience achieved in the Hungarian and South Transdanubian small towns. Raitano and Marcoaldi consider that, although the implementation of their master plan is still in progress, the projections have not been achieved so far. Leonidakis is also pessimistic about the situation in Naoussa since most of the local industries are still inoperative and their buildings are vacant. Ramé believes that although the initiative was good in the BioComechingones Corridor, implementation has been unsuccessful. Pellegrini is pessimistic in the case of the Stella river valley since the research results have not been converted into action on the territory, while on the border between Prepotto and Slovenia a certain incipient renewal can be observed in the fabric of gastronomic production and heritage.

Related to the above, one of the keys to resilience is based on the achievement of a situation of balance that, in addition to success at the time of the implementation of projects or initiatives, should be accompanied by its continuity over time. In this sense, and given the authors' opinions, there is a diverse perspective in terms of forecasting the durability of actions in the future. Some contributors are generally optimistic, as is the case with Royo, since the Valle del Guadalhorce agricultural park should be consolidated in the future, maintaining promotion and education, as Ostos and Aladro believe about Granadilla, with the continuity of summer schools, although alternative activities such as leisure or sports should be proposed in their opinion. Gonçalves, Rosendahl and Pérez expect a gradual and continuous growth of resilience in Cernache do Bonjardim, although some time is needed to verify this. Janáč is also optimistic about the apparent continuity of good results at Vysoké nad Jizerou. Ramé believes that the continuity of the project in the BioComechingones Corridor will be subject to official support, either from the national or provincial administrations, while Astrella is enthusiastic about the growing success of the project in the Avelino-Rocchetta line, accompanied by new initiatives that have emerged in the first years of its implementation. Janáč, Elts and Huovinen make no specific mention in the cases of Telč, Bardejov, Ulyanovsk and Salento respectively. Among those who are sceptical about durability, we can mention Horeczki, who sees no clear signs of continuity in the Hungarian and South Transdanubian small towns, or Raitano and Marcoaldi on Viterbo, who think that it is necessary to develop and maintain two lines of work on a still incipient project; the city lab and two strategic interventions that are already programed. Leonidakis, on Naoussa, shares his doubts about the uncertain future of a process that is not yet finished. Pellegrini is pessimistic about the cases of the Stella river valley and the Slovenian border with Prepotto since the first is apparently stuck, while in the second no initiatives have appeared beyond the implementation of the studied strategy.

As for the materialization of the results obtained in each of the cases, we will start with those that have affected the physical state and integrity of the goods, that is, on the heritage conservation, in many cases linked to the idea of marketing of the historical architecture without without any awareness of what it is on the side on the owners. In most small towns, there has been a direct impact in this field, as in Avesta and Söderhamn, as a result of the transformation of former industries into service centres. In the Guadalhorce Valley, the focus has been on the rehabilitation and conservation of traditional architecture, more specifically the cascareros, while in Viterbo it has been one of the priorities of the master plan, both in the town walls and in the architecture and the whole urban fabric. In Naoussa the effect has only reached some elements of the built heritage, leaving a large part of it still in disuse and without being rehabilitated. In Granadilla, the reconstruction of various ruined buildings has been one of the main actions, although most of the walled enclosure has yet to be rehabilitated. In Cernache do Bonjardim, conservation has been more focused on natural heritage and, if anything, on vernacular architecture. In the BioComechingones Corridor investments have focused on the regeneration of buildings and public spaces for tourism, which has also been the case on the Stella River, while in Prepotto interventions have been introduced for local producers. In Avelino-Rocchetta efforts have focused on the conservation of the railway infrastructure, including its restoration in some points, as well as that of the historic trains. In Salento, most of the conservation actions have been linked to the built heritage and, in some cases, to the commodification of local traditions. In Telč the material heritage conservation has been guaranteed by the promotion and facilities for tourism, while in Bardejov it has not been something generalized, with problems remaining in the Jewish suburb. In Ulyanovsk, urban and architectural regeneration, although not one of the priorities of the initiative, has been implicit in cultural policies and programs. The direct impact on tangible heritage conservation has also been limited in the Hungarian and South Transdanubian small towns, where only a few buildings housing public uses have occurred, as most of the built heritage was destroyed by the war. In Vysoké nad Jizerou, by contrast, the conservation of the built heritage was not a priority, but initiatives have been more oriented towards the intangible heritage, except for the reconstruction of a historic town gate.

Among the objectives not related to material conservation, but focused on the social sphere, one of the priorities common to rural areas, small towns and their territories is, according to scientific literature and many of the experiences, the anchoring of the population, in contexts marked by national and international migration towards urban centres and places with better prospects for economic development and employability. However, in many of the cases collected in this volume, such as the case of the Guadalhorce Valley, Salento, Telč and Bardejov, the anchoring of the population has not been one of the priorities of the initiatives carried out. Nor has it been so in Naoussa, where external and internal migrations have taken place, nor in the territory of Friuli Venezia Giulia, where, moreover, the lack of critical mass has been one of the main problems. In Avelino-Rocchetta, although it is not mentioned as one of the objectives, it has probably been a positive side effect, just as it has been in Ulyanovsk, while in Granadilla it is not one of the possibilities or of the initiatives that make sense at the moment. In Viterbo, although the anchoring of the population is not discarded in the future, it is not one of the results at the moment within the historical centre. On the contrary, there are cases in which this has been one of the priority aims of the actions undertaken, such as in Cernache do Bonjardim, where the anchoring of the local population is still fundamental, and likewise, it is one of the pillars of the strategy in Vysoké nad Jizerou, and in Avesta and Söderhamn, in which gaining population was a strategic objective and still is with positive results, largely due to immigration and the arrival of refugees. In the Hungarian and South Transdanubian small towns, although this was one of the supposed objectives of the initiatives launched, no specific measures are being taken at the moment. In the BioComechingones Corridor, this has been one of the main objects of study, although the arrival of external population entails a risk of loss of local population and identity.

Finally, we will analyze two aspects related to social resilience in some of the case studies addressed in the book. The first of these refers to the presence or absence of an increase in community resilience as a result of the actions taken. In many of the cases, this has not been considered a relevant outcome, although there have been exceptions. The clearest cases are those where this was already considered one of the fundamental objectives in the conception and implementation of the initiatives, such as in Cernache do Bonjardim, where successful results have been obtained, and in Vysoké nad Jizerou, where, as one of the pillars of the project, it has been achieved. In other cases, such as Viterbo, although it was included among the principles of the master plan, it has not been very successful, mainly due to the lack of interest on the part of the citizens. In the Avelino-Rocchetta line, the sense of community was only maintained in the first phases of the project, fading away later on, while in Granadilla, although there is a sense of resilience based on the local identity of the former inhabitants, there is no real implementation on the ground. In Corredor BioComechingones, the development of the project itself puts at risk, precisely, the community resilience.

The second aspect is even more residual in terms of the social and economic impact of the initiatives studied in the volume and is related to a question that, according to the literature, is generally associated with the processes of heritage commodification. This concerns touristification and, beyond that, the effects of over-tourism on heritage sites. In the case of the small towns studied, this does not seem to be a problem that has affected resilience in a generalized way. In most cases, it is not expected, while in some, such as Friuli Venezia Giulia, Granadilla or Ulyanovsk, it is an improbable risk. It also seems unlikely in Vysoké nad Jizerou, given the strong local resistance to changes in its social structure and identity. There are only three cases in which tourism seems to be a probable threat: in the BioComechingones Corridor the risk already exists, and it is actually one of the main menaces to the stability and resilience of the existing social and heritage fabric, introducing the phenomenon of post-tourism into this book. In Telč and Bardejov it is also considered a major risk, and very high in the case of Telč since this small town is already full in terms of tourism capacity. The problem is similar, or even more pressing, in Salento, being probably the aspect most intensively dealt with in this case study.

Some keys on small towns resilience and heritage commodification

This book shows a varied collection of small towns where strategies based on heritage commodification have contributed to their resilience. Among the case studies chosen by the different authors, we distinguish urban and rural areas, small, medium and large scale intervention areas, and different types of initiatives – based on cooperation, the local community or urban and territorial heritage resources, among others – characterized by a very diverse heritage, both tangible and intangible. We can, therefore, say that small towns offer a sufficiently broad base to develop all kinds of actions aimed at resilience through heritage commodification.

Concerning the initiatives and projects presented in the different chapters, these have, with a couple of exceptions, a top-down approach, led by the authorities at a local or regional level, although sometimes there is collaboration at a national level for strategic reasons, or even at an international or cross-border level. This starting point determines, to a great extent, certain steps when defining and promoting the strategies established for the heritagization and commodification of resources in each case. We are referring, in general, to the attraction of public investment through international calls, and the promotion from the institutions themselves of strategic projects in the different territories with their budget. We can find also cooperation projects between administrations at different levels or between institutions in nearby locations to join forces. All of them are accompanied by the establishment of guidelines, plans, projects and official documents that have tried to regulate different aspects of these towns, such as population growth, the rehabilitation of historical areas, the promotion of tourism and the strengthening of a certain aspect of local cultural identity.

The relationship with the local population was, however, one of the revelations when analysing the set of case studies selected in the book. Although in two of them the initiative has clearly been grass roots, and therefore the work with the community has been fundamental, in most of the projects there is no special emphasis on achieving community building or even any clear involvement of the inhabitants of the towns within the design and implementation of the actions carried out. In the cases in which this relationship does exist, there is certain independence from large public or even private investments, as they tend to generate strategies based on the circular or collaborative economy, with small contributions from individuals based, fundamentally, on the feeling of belonging to the place and the reaffirmation of local identity.

This is also the case in those areas where tourism promotion has been one of the fundamental objectives, since owners and service providers have made small investments to start up their businesses, which, with or without public subsidies, has contributed to the progress of the joint initiative. Indeed, the promotion of heritage seems to be one of the fundamental objectives of the different projects studied, whether they are aimed at the external market or, less usually, at the inhabitants themselves, which is directly related to commodification. In many cases, the outside promotion has been aimed at seeking the tourist use of heritage resources, both within the towns and in strategies on a territorial scale. Only in some cases has it been accompanied by an attempt to combat population shrinkage, either by promoting the dynamism of these towns to attract new inhabitants or by anchoring the existing population in them. In certain cases, the internal promotion strategy has been accompanied by cultural or educational programs, although this has not been the general trend observed.

These issues have had a direct impact on the results obtained by each of the initiatives addressed in this book. In this sense, community resilience has been an outcome in few cases, and in fact, has been compromised as a result of other initiatives that have put too much strain on the balance between local people and over-tourism, especially when the anchoring of the population has not been an objective accompanying the strategies of heritage dissemination. However, tourism seems to be an unlikely risk in most of the cases studied, which distances this collective study from others carried out about capitals, metropolitan areas or large heritage centres, and contributes to the potential for the resilience of small towns in terms of the implementation of heritage commodification strategies.

On the contrary, most of the projects studied tend to obtain results in the material aspect, that is, in the conservation of heritage assets, mainly immovable, which constitute the resources on which the commodification strategy has been built. Most investments have not only resulted in planning but also in motivating direct intervention on buildings, urban and peri-urban spaces, including heritage infrastructures and historical agricultural resources on a territorial scale. In a few cases, conservation has been oriented towards the intangible heritage, local identity or culture.

Finally, the results obtained by the different projects must be assessed to know to what extent the strategies of heritagization and commodification have contributed to the resilience of the small towns considered as case studies. In this sense, and even though most authors are optimistic in the short term regarding the achievement of objectives, they are also prudent when it comes to evaluating their continuity and effective success in the long term, with few exceptions apart, perhaps, from those who have worked on case studies based on community resilience. They generally have more confidence in their persistence. In some cases, we can state how heritage-based strategies cannot stop or change other more powerful economic and demographic forces, such as the dynamics related to demography, economy or the central place function that may cause a big disturbance in the geographic equilibrium. Many of them concur that the agreements, projects and strategies analysed in the different chapters are still in a phase of recent approval, or in the first moments of their implementation, and some of them have already acquired stagnant patterns which have not had the expected impact in the first years of development or whose formalization into effective actions on the territory has not come about. Therefore, it will be necessary in the future to continue evaluating the progression of these and other examples of resilience strategies in small towns based on heritage commodification, as well as to deepen and broaden studies in a field that, through this collection of contributions, demonstrates its potential for a more sustainable and balanced future based on local resources, collaboration and culture.

Acknowledgments

This work has been partially supported by the project "SIN_PAR: Sistema de Innovación para el Patrimonio de la Andalucía Rural", funded by European Regional Development Fund (FEDER) - Consejería de Economía, Conocimiento, Empresas y Universidad de la Junta de Andalucía - Instituto Andaluz del Patrimonio Histórico (ref. no.: PY20_00298), which will allow the various authors of this book to to further develop research about heritage-based strategies for local development in small, remote and rural places.

This work was supported by the European Regional Development Fund project "Creativity and Adaptability as Conditions of the Success of Europe in an Interrelated World" (reg. no.: CZ.02.1.01/0.0/0.0/16_019/0000734) implemented at Charles University, Faculty of Arts. The project is carried out under the ERDF Call "Excellent Research" and its output is aimed at employees of research organizations and Ph.D. students.

Note on authors

José Manuel Aladro Prieto is an Associate professor in the department of Architectural History, Theory and Composition, ETSA, University of Seville, Spain, since 1999. Architect, Ph.D. in Architecture (2012) by the University of Seville. Collaboration in Master and Ph.D. programmes in Architecture and Heritage. Deputy Secretary of the Master's degree in Architecture and Historical Heritage of the University of Seville (2006–2018). Director of the department of Architectural History, Theory and Composition, ETSA (2021 onwards). Research award "Best Industrial Heritage Research in Andalusia" by the Fundación Patrimonio Industrial en Andalucía (2013). Award "Architecture Promotion", by Colegio de Arquitectos de Cádiz in 2001–02, 2007–08 and 2012–13. His research lines include industrial architecture and wine-related cityscape and landscape. Author of "The urban scale of Jerez. Sherry winery complexes in the 19th century" (2021). Co-author, among other publications, of "Fortifications and defence systems guide of the Bay of Cadiz" (2012) and "Farmhouses, Farms and Winepresses. Architecture of agricultural settlements in Andalusia, Province of Cadiz" (2002).

Consuelo Isabel Astrella is an Architect, Ph. D. in Architecture and a freelance researcher. She has been a post-doctoral research fellow at the Department of Architecture of (DiARC) the University Federico II of Naples. She is interested in the field of Heritage and Architectural Conservation also applied to the outskirts of Italian cities as in a recent research project to which she has taken part. She graduated in 2015 with a master's degree in Architecture at DiARC of University Federico II of Naples with a thesis in Theories and History of Restoration. She became Ph.D. in Architecture in 2019, at the same university, with a research thesis about the abandoned railways as a complex heritage of industrial archaeology to preserve and enhance, with their territories and landscapes. She published some scientific articles and contributions on these topics and, during her studies, she deepened her knowledge about railway heritage and conservation thanks to a few months visiting period in 2018 at the University of York and the NRM of York (UK).

Elena Elts is a specialist in the field of foreign cultural policy, regional cultural policy, international museum cooperation, protection of cultural heritage and regularly participates in the ICOM conferences. Author of more than 40 articles and methodological papers in Russian and English. E. Elts has a membership in Saint-Petersburg Association of International Cooperation. E. Elts is engaged in research of regional public diplomacy and international cooperation in the field of preservation of cultural and natural heritage and protection of cultural rights. Diploma of Specialist in Art History, Saint Petersburg State University, with honors, June 2004. Ph.D. in history, January 2008. 2010 and now assistant professor of Department of International Cultural and Humanitarian Cooperation, School of International relations, Saint Petersburg State University.

Blanca Del Espino Hidalgo is an Architect, Ph.D. in Architecture, Master in Architecture and Historical Heritage, Master in Sustainable City and Architecture. She works as a Doctor Researcher in the Andalusian Institute of Historical Heritage, Area of Built Heritage, Department of Projects, and since May 2021 she is the Head of the Centre of Documentation and Studies in the same institution. She is also a researcher at the University of Seville (Escuela Técnica Superior de Arquitectura), linked to the research group HUM-700 "Heritage and Urban Territorial Development in Andalusia". She has developed her main line of research on the relationship between urban-territorial heritage and sustainability, including the study of medium-sized cities, rural areas and small towns, and metropolitan areas. Additionally, she develops studies on connected areas such as tourism, landscape, citizen participation or conservation of built heritage. She has carried out five research stays abroad in Europe, has published three monographs and numerous papers in prestigious scientific journals and publishers. She has supervised three doctoral theses and ten master theses.

Marta Marçal Gonçalves is Dr. Engineer. Full time Assistant Professor at the Civil Engineer Department of the University of Algarve, Portugal, since 1994. Civil engineer (Instituto Superior Técnico, Lisbon, Portugal), Operational Research specialist (University of Lisbon, Portugal) and Ph.D. in Urban Planning and Territorial Planning (University of Seville, Spain). Has supervised several final works of the Bachelor's Degree in Civil Engineering of the University of Algarve. She was part of the "Tuna

Route, a bluefin tuna experience" project team. Collaborating member of the research group HUM-700 "Heritage and Urban and Territorial Development in Andalusia". Currently its line of research integrates themes on resilience and vernacular heritage, focusing on dry stone walls and tuna fishing.

Réka Horeczki is a junior researcher in Institute for Regional Studies CERS Hungarian Academy of Science. Economist, Ph.D. in Economics (2020) by the University of Pécs. Her research areas have focused on local and regional development processes, especially on the socio-economic development of small towns in Central and Eastern Europe; futhermore the measurement possibilities of the urban-rural dichotomy and of liveable small-town (with social-cultural attributes, long-term urban processes etc.). She collaborates and prepare in national and international projects and researches.

Katja Maaria Huovinen received a master's degree in Architecture at Helsinki University of Technology in 1996. In her licentiate thesis (2009) she has studied the dialogue between past and present in a Southern Italian agrotown. In her Sc.D (2017) she has studied the practice of urban conservation in three European World Heritage sites. Specialized in history, urban history and conservation, she has been running her own architectural studio since 2006. Besides her private practice, she is doing independent research.

Jira Janáč is Assistant Professor at the Seminar of General and Comparative History, Institute of the World History at the Faculty of Arts and a post doctoral researcher at the Academy of Science of Czech Republic. He has graduated cum Laude in the Ph.D. programme under double supervision at the Charles University, and the Eindhoven University of Technology in the Netherlands. His research interests focus on the social and cultural history of technology and its methodology, heritagization of technologies in European identities, environmental history and heritage studies.

Luda Klusakova worked at the Faculty of Arts, Institute of World History, Seminar of General and Comparative History Charles University, for more than 40 years. She graduated from two programmes, History and French, at CUFA. Briefly, she worked as a research assistant at Archive of Charles University; from 1978 until the present, she worked at CUFA. In 2000, she became the director of Seminary of World and Comparative History. She was expertly focused on the comparative study of cities' modernization, their image in culture, and the use of cultural heritage and history as a strategy of local and regional development. She contributed to the Institute's engagement in eminent European research and educational projects, for example, international MA programme TEMA+ (European Territories: Heritage and Development). The French Government awarded her with the Order of Merit in Education, Science and Culture Chevalier de l'ordre des palmes académiques. Between 2010–2012, she was the President of the European Association for Urban History. In 2019, she was appointed the Foreign Member of Finnish Academy of Sciences for her exceptional scientific merits.

Jan Krajíček was born in 1986 in a small town in the Bohemian-Moravian Highlands, maybe therefore long-time interested in historical development of peripheral regions. MA degree in General and Comparative History (CUNI). Previous research on Central European distant localities, development projects and the use of local cultural heritage. Working in REACH project (Redesigning Access to Cultural Heritage) and Czech Ministry of Social Affairs Project Místa Zblízka (support of local social participation using cultural heritage). Currently finishing his Ph.D. on the comparison of development strategies of post-socialist European regions. Keen motorbiker (Jawa Pérák, 1953) and occasional moto-journalist.

Nikolaos Leonidakis was born in Athens, Greece, in 1972. He is a doctoral student in Modern and Contemporary History, at the History-Archaeology Department of University of Crete. The area of his studies is the Non Performing Greek Business Loans in the 1950s, within the socioeconomic and political framework of the era, and his scientific interests are focused on post war labor and economic history. He holds a Master's Degree in Modern History from University of Crete. His thesis was about the post Greek Civil War's financing of a "bandit- stricken" Wool Spinning Mill, with Marshall Plan's funds. He has a Bachelor's Degree in Greek Philology from Aristotle University of Thessaloniki. He is a member of the Greek Economic History Association and the Greek Oral History Association. He has presented many papers in various Greek and international History conferences.

Paolo Marcoaldi is architect and Ph.D. in Architectural Design and Theory. Since 2011 he has been teaching and researching at the Faculty of Architecture of Sapienza University of Rome. Since 2013 he is redactor of the ArchiDiAP web portal. His research is focused on: Architecture of modern Rome, methodological issues in architectural design. In 2017, with Fabio Balducci, he founded "Dismisura", an architectural practice based in Rome.

Lars Nilsson is Professor emeritus of Urban History and former Head of the Institute of Urban History at Stockholm University. He was President of European Association for Urban History 2004–2006 and member of the Association's International Committee 1998–2012. He is a member of the International Commission for the History of Towns since 1994. Among his publications, the following could be highlighted: Svensk stadshistoria – En introduktion (2019 co-authors Mats Berglund and Per Gunnar Sidén), The democratization of the Swedish Town Hall in the early 20th century, Political functions of urban spaces and town types through the ages, Roman Czaja et al. (2019), The Post-Industrial City: Main trends in European Urban Growth, 1970–2015, Proceedings from the International Conference "Cities through History" Guimaraes 2017 (2018), Krympande orter: Avesta och Söderhamn som postindustriella samhällen (2016, co-author Eric Båve), The Coming of the Post-Industrial City (editor and author, 2011).

Francisco Javier Ostos Prieto is an Architect (2017) and research fellow (PhD) in the Department of Urban and Land Planning of the High Technical School of Architecture, University of Seville, within the research group HUM-700 "Heritage and Territorial Urban Development in Andalusia". Master in Architecture and Historical Heritage and Master in Urbanism, Planning and Urban Design. Both with works developed on convent heritage, one titled "The convent structure in Écija. Genesis, influences, transformations and continuities in medium-sized cities" and the other "The convent foundation in the Andalusian context. Territory, city and urban protection". He has obtained the Jóvenes Líderes Iberoamericanos Scholarship (2018) by the Fundación Carolina and the DAAD Scholarship (2019) for studies in Germany by the Deutsche Akademische Austauschdienst. He has also participated in numerous conferences and articles in journals related to his research, highlighting the main one called "Identity construction of the European medium-sized city through the monasticism repercussions in Écija" (2019).

Paola Pellegrini is Ph.D. in City Planning and Urban Design and Associate Professor of Urban Design at the Department of Urban Planning and Design of XJTLU. She was Visiting Associate in Urban Planning and Design at the Harvard Graduate School of Design, Cambridge MA; adjunct Associate Professor at the European postgraduate Master in Urbanism – Consortium of TU Delft, KU Leuven, UPC Barcelona, IUAV Venezia – and at the Politecnico di Milano in Italy. Her research activity focuses on urban heritage and urbanization and adopts a research-by-design approach in exploring the contemporary urban phenomena both in Europe and in China. Paola is also a licensed Architect and has often combined academic research and professional practice, recently for European Projects related to urban design and mobility infrastructures.

María Teresa Pérez Cano is Dr. Architect. Full time professor at the Urbanism and Territory Planning Department of the University of Seville, since 1995. Permanently collaborating with several, Master's and has been already supervisor of more than 30 doctoral thesis. Co-founder and director of the research group HUM-700 "Heritage and Urban and Territorial Development in Andalusia". She has led a R&D project on medium-sized cities in Andalusia. Focused in urban planning preservation, the group has conducted several researches, urban plans and professional advisory on classed buildings, monuments, historic centers and landscapes. She has taught on Doctorate Programs in the University of Seville, Universidade Moderna (Lisbon and Oporto), University of Zulia, Maracaibo (Venezuela), Central University of Santiago de Chile (Chile), PUC-Campinas of Sao Paulo (Brasil), the Architects' Association of Málaga, Huelva, Córdoba, Badajoz, etc.

Manuela Raitano is Associate Professor at the Department of Architecture and Project of Sapienza, University of Rome, where she teaches at the Ph.D. in Architecture and Construction. She also conducts research activities at the "Architecture and Context Laboratory" (ArCo) and is the editor-in-chief of "ArchiDiAP – condividere l'architettura / ArchiDiAP – sharing architecture", a web portal dedicated to the scientific dissemination of modern and contemporary architecture. Her writings and projects are published in national and international journals and exhibition catalogues.

Mónica Jimena Ramé got her degree in Architecture at the Faculty of Architecture, Urbanism and Design -FAUD- from the National University of Córdoba -UNC-. From 2013 on she is assistant professor at the chair of Latin American Architectural and Urban History - FAUD- and a scholar at the National Research Council (CONICET) at the binational doctoral program on Urban Regional Studies -Bauhaus Universität Weimar, Germany -UNC, Argentina. She contributes with the blog "Un Día Una Arquitecta" and the Virtual Museum of Architecture of the University Museum Program UNC. Her research interests cover the issues of history of Latin American architecture and urbanism with a gender perspective, inclusive and equitable public spaces design and mountain regions sustainable urban planning.

Stefan Rosendahl is Ph.D. in Natural Sciences. Born in 1956, Diplom-Geologist and Ph.D. in Natural Sciences since 1985 (Geology and Paleontology, University of Stuttgart, Germany). Full time professor and head of Civil Engineering Department at the Universidade da Beira Interior (Covilhã, Portugal) until 1991. Full time professor at the Higher School of Art and Design of Caldas da Rainha (Portugal), until 1996. Full time Professor at the Universidade Lusófona de Humanidades e Tecnologias (Lisbon, Portugal), with activities in Ecodesign and Bionics for Design, until 2018. Actually, professor at the Instituto Superior D. Dinis (Marinha Grande, Portugal) with lessons of Ecodesign and Bionics for Design. Current line of research consists in activities and studies about geotourism, cultural tourism and natural heritage.

Lourdes Royo Naranjo is Full Professor at the University of Seville, in the Department of History, Theory and Architectural Composition of the Superior Technical School of Architecture since 2005. Director of the research group HUM-1050 "Heritage Action Group". After completing her doctoral thesis in 2010 with a FPU grant from the Ministry of Education and Science, during this time her research shows the ability to establish interdisciplinary relationships between culture, art and architecture of the 20th century. Degree in art history and Bachelor of Heritage Conservation and Restoration. Master in Architecture and Historical Heritage and Higher Master in Heritage Resources, she has carried out different research stays in Universities such as the Technical

University of Lisbon and the University of Algarve, both in Portugal, the Università degli Studi Roma Tre (Italy), the Universitat Poltiècnica de Catalunya and the University of Malaga. Likewise, she has participated in numerous International Congresses on urban planning and spatial planning, tourism, heritage and architecture of the 20th century.

Europe of Cultures

Dialogues

"Europe of Cultures" is a series of studies, monographs, stories, research projects, reports on conferences and debates devoted to the complexities and changing realities in European societies. It bridges the past with the future at the cross road of challenges and opportunities of the transformation of European societies. The management of changes in societies refers to the interconnection of various dimensions and levels of policy-making impacting on economic, social, political, democratic, communication, philosophical, artistic, religious as well as ethical traditions and behaviour. As an editorial project the series is structured along two interconnected and complementary sub-series: i.e. the "(Europe of) Dialogues" series and the "(Europe of) Living Stories".

The sub-series "(Europe of) Dialogues" mainly deals with (cultural) diversities, identity and citizenship building in Europe as well as with the relevant multi-level governance and communication structures in the transformation of European societies. Europe is a laboratory for understanding this multi- and intercultural reality. The purpose is to contribute to a better understanding and communication of the changes taking place by looking at the European societies in general, and the specificities of different national, regional and local cultures and communities in a framework of dialogues. The series presents interdisciplinary and critical views of value-driven and policy-oriented reflections. Moreover, it offers new insights into understanding how to manage, value and communicate cultural diversity, identity and citizenship. It also wants to contribute to the development of new ways of "living together", in which cultures and communities are perceived as binding forces in creative society building.

The sub-series "(Europe of) Living Stories" (formerly "Mémoire de l'Europe en devenir", Director Gabriel Fragnière †) is devoted to inspiring narratives for a broad public with a view to contribute to a better understanding, communication and contextualisation of the newly emerging

Europe. It mainly focusses on stories, memories and testimonies of persons, events, institutions and issues that have transformed mentalities, fostered European awareness and finally shaped Europe's future. These stories serve as important references and communication tools for future developments of Europe in the world. This collection wants to be open and diverse, original and dynamic in its content, method and pedagogy faithful to Europe's role and reference in the globalising world.

Series Editor : Léonce Bekemans,
Jean Monnet Professor ad personam, Bruges, Belgium

Editorial Board

Series titles

Vol. 1. Mark Dubrulle et Gabriel Fragnière (dir.), *Identités culturelles et citoyenneté européenne. Diversité et unité dans la construction démocratique de l'Europe*, 2009.

Vol. 2. Gily Coene et Chia Longman (dir.), *Féminisme et multiculturalisme. Les paradoxes du débat*, 2010.

Vol. 3. Muriel Rouyer, Catherine de Wrangel, Emmanuelle Bousquet et Stefania Cubeddu (dir.), *Regards sur le cosmopolitisme européen. Frontières et identités*, 2011.

Vol. 4. Lénia Marques, Maria Sofia Pimentel Biscaia and Glória Bastos (eds.), *Intercultural Crossings. Conflict, Memory and Identity*, 2012.

Vol. 5. Léonce Bekemans (ed.), *A Value-Driven European Future*, 2012.

Vol. 6. Albert Doja, *Invitation au terrain. Mémoire personnel de la construction du projet socio-anthropologique*, 2013.

Vol. 7. Alain Michel, *Dompter le dragon nucléaire. Réalités, fantasmes et émotions dans la culture populaire*, 2013.

Vol. 8. Jacques Migozzi, Stéphanie Delneste et Olivier Odaert, *Les racines populaires de la culture européenne*, 2013.

Vol. 9. Laure Clément-Wilz et Sylvaine Poillot-Peruzzetto (dir.), *Construire la citoyenneté européenne*, 2014.

Vol. 10. Pascaline Gaborit (ed.), *The Strength of Culture for Development. Why Culture Matters in International Cooperation*, 2014.

Vol. 11. Albin Wagener, *L'Échec culturel. Vie et mort d'un concept en sciences sociales*, 2015.

Vol. 12. Céline Romainville, *European Law and Cultural Policies / Droit européen et politiques culturelles*, 2015.

Vol. 13. Bertrand Vayssière, *Penser les frontières européennes au XXIe siècle*, 2015.

Vol. 14. Monica Simeoni (ed.), *Europe or Not ! Multiple Conversations and Voices. With Alberto Martinelli, Vittorio Cotesta, Nadia Urbinati and Alain Touraine*, 2016.

Vol. 15. Michel Theys, *Jacques-René Rabier. Fonctionnaire-militant au service d'une… certaine idée de l'Europe*, 2017.

Vol. 16. Paal J. Frisvold, *Towards Europe. The Story of a Reluctant Norway*, 2018.

Vol. 17. Sabine Menu, *Paul Collowald, pionnier d'une Europe à unir. Une vie à dépasser les frontières*. 2018.

Vol. 18. Emery-Pamphile Mbadu Mbadu, *Reconstruire l'Europe avec Jean-Marc Ferry. Perspectives d'un État cosmopolitique*. 2020.

Vol. 19. Cesáreo Rodríguez-Aguilera de Prat, *A United States of Europe?*. 2020

Vol. 20. Juan Prat y Coll, *Mes séjours à Bruxelles. Témoignages diplomatiques* (1986–2020). 2021.

Vol. 21. Cornelia Constantin, *Les pères de l'Europe en héritages*. 2021.

Vol. 22. Maddalena Colombo, *Intercultural Issues and Concepts. A Multi-Disciplinary Glossary*. 2021.

Vol. 23. Luďa Klusáková and Blanca Del Espino Hidalgo (eds.), *Small Town Resilience and Heritage Commodification*. 2021.

www.peterlang.com

www.ingramcontent.com/pod-product-compliance
Ingram Content Group UK Ltd.
Pitfield, Milton Keynes, MK11 3LW, UK
UKHW041834100926
8556IPUK00018B/956

9 782807 617438